Houghton Mifflin Books

by ISAAC ASIMOV

Breakthroughs in Science

The Greeks: A Great Adventure

The Kite that Won the Revolution

Quick and Easy Math

Realm of Algebra

Realm of Measure

Realm of Numbers

Words from the Exodus

Words from the Myths

Words in Genesis

Words of Science

Words on the Map

LIBRARY OF CONGRESS CATALOG CARD NUMBER 65-19294

PRINTED IN THE U. S. A.

Second Printing R

Isaac Asimov

An Easy Introduction to the Slide Rule

with Diagrams by
William Barss

Houghton Mifflin Company Boston
The Riverside Press Cambridge

CONTENTS

To a friend
who has accompanied
and served me
loyally for
twenty years

My Slide Rule

1

Arithmetic by Distance

To Begin With—

WE HAVE all heard, these days, of electronic computers. These marvelous instruments, which came into use during World War II, are capable of performing in a few seconds work that might take years if all we could use were pen and paper.

There are times when arithmetical problems come our way and we might wish that we ourselves owned such a computer to do the work for us. Such a situation would have its disadvantages, however. Electronic computers are bulky, expensive, complicated, and can be handled only by people with special training.

Besides, electronic computers aren't at their best when used for everyday problems. That would be like trying to shoot a fly with naval artillery.

For a fly, an ordinary swatter is much better, and for ordinary mathematical problems, we could best use a really simple computer.

There happens to be a simple computer, just suitable for everyday computations, that was invented about 350 years ago. It isn't electronic; there are no electric currents involved. In fact, it is no more than a piece of wood with some marks on it. It looks like a ruler except that it has a middle piece that can slide back and forth, so that it is called a *slide rule*.

A slide rule doesn't seem as impressive as a giant electronic computer, but it has many advantages. It is small enough to put in your pocket, it need not cost more than a couple of dollars, it can't go out of order, and, best of all, it can solve almost any numerical problem that you meet up with under ordinary circumstances.

To add to all that, it is simple to operate. If you know grade-school arithmetic, you can use a slide rule, even though you may not quite see why it works! If you have taken some high-school mathematics also, the reason for its success can be explained with very little trouble.

In this book, I shall start from the beginning and try to show you both how and why it works.

To be sure, merely reading this book will not make you an expert at handling the slide rule. For that you will need practice and I am not including practice exercises as part of this book.

However, once you thoroughly understand the principles of the slide rule — once you know what you are doing and why — then it will be simple to set yourself problems. It will be simple for you to use the slide rule on problems that arise from day to day.

As you practice you will become expert, and you will be amazed to see how, by merely sliding one piece of marked wood against another, computations that seem very complicated can be completed in a few moments.

In fact, if the time comes when you have a job in which numerical computations have to be carried through frequently, you will want your slide rule with you at all times. Without it, you would feel like a doctor without his stethoscope or a painter without his brush.

Let's consider, then, how a piece of marked wood can help us in our calculations.

One Ruler

One of the earliest tools used by civilized men engaged in fine and accurate work — as in making plans for building temples or tombs — must have been a piece of wood or ivory or metal which had a straight edge. By sliding a pencil, or any marking device, down this straight edge, a straight line could be drawn.

You understand how important it is, in drawing up any accurate plan, to produce perfectly straight lines, but to do so some guide is absolutely necessary. Try drawing straight lines freehand, that is, without using some straight edge to guide you, and see what a sloppy appearance it makes.

An instrument possessing a straight edge and nothing more is called just that, a *straightedge.*

It is simple, however, to put the straightedge to another use, and make it a means for determining the length of a straight line.

For this purpose it is only necessary to make small

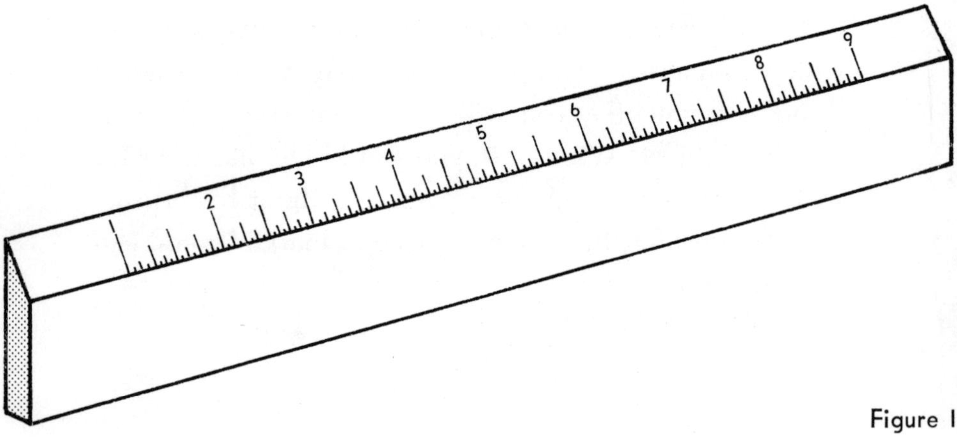

Figure 1

marks upon it — marks spaced a fixed difference apart. Such a marked straightedge is a *rule* or a *ruler*.

The ruler with which we are most familiar has the marks upon it spaced an inch apart, and is 12 inches long all together. Since 12 inches make 1 foot, such a ruler is a *foot-rule*.

The foot-rule has its inches subdivided into halves and quarters, usually eighths as well, and sometimes even into sixteenths (see Figure 1).

Such a ruler can be used to determine the length of a line or to draw a line of a particular length, and we are all familiar with its use.

Can we, however, use it for some other purpose than measuring length? Suppose we think about that a bit . . .

When we say that $1 + 1 = 2$, we are talking about "pure numbers." The 1's and the 2 in that sum do not represent one of anything, or two of anything. Still the sum can be made to apply to numbers that are not pure — to count objects.

For instance, 1 apple + 1 apple = 2 apples; 1 chair + 1 chair = 2 chairs; 1 star + 1 star = 2 stars; and 1 inch + 1 inch = 2 inches. We can say this not only for a sum like 1 + 1 = 2, but for any arithmetical problem that involves addition or subtraction. We know that 72 + 28 = 100 and that 35 − 20 = 15, and we can be sure, therefore, that 72 inches + 28 inches = 100 inches, and that 35 inches − 20 inches = 15 inches.

In other words, if we add inches to inches, or subtract inches from inches, the numerical portion of the answer we get would be the same as the numerical portion we would have obtained if we had used apples, chairs, stars, or anything else. It would be the same, indeed, as answers we would have gotten in using pure numbers.

This means that if we could somehow use some device to prove that 45 inches and 32 inches taken together make up a length of 77 inches, then we can be sure that 45 + 32 = 77. The device we would use to put lengths together in this manner would, in effect, do our addition for us.

Let's take a very simple example. Look at a ruler (or, if you don't have one handy, look at Figure 1) and, starting at the left edge, count off 2 inches with your finger. Your finger is now pointing to the number 2. From there, count off 3 more inches. Now your finger is pointing to the 5. What you have done is shown yourself that 2 inches + 3 inches = 5 inches, and you can deduce from that that 2 + 3 = 5.

Try again. Count off 3 inches and, starting from the place you reached, count off 5 more, and you will find that 3 + 5 = 8. In the same way you can use an ordinary foot-rule to prove to yourself that 4 + 7 = 11, 1 + 8 = 9, or even that 6 + 3 = 3 + 6, for both these sums come out to 9.

Subtraction works just as well. Count off 10 inches and you are at the number 10. If you want to subtract 3 inches from that, count them backward, chopping each inch off the total length. When you finish you find yourself at the number 7. This means 10 inches − 3 inches = 7 inches, or 10 − 3 = 7.

What we are doing is manipulating lengths and using those lengths to tell us something about numbers. Lengths are not numbers, of course, but in certain ways they follow the same rules that numbers do. Lengths have properties that are *analogous* to those of numbers in some ways. If we then use lengths to work out, or "compute," the answers to number problems, we are using the foot rule as an *analog computer.*

An example of a very elaborate analog computer is Univac, which is used by television studios to work out the course of elections in progress and to predict the winner. (Univac stands for *Univ*ersal *A*nalog *C*omputer.)

Univac makes use of flashing electric currents that alter the position of many thousands of tiny switches in millionths of a second. The properties of the changing switch positions are analogous to those of numbers, which is why Univac can be used to solve problems. However, an ordinary ruler, although much, much simpler than Univac, can also be used as an analog computer, and we've just done it. In fact, it is easier to determine that $2 + 3 = 5$ on a foot-rule than it is on Univac.

Two Rulers

Next, let's see how we can make a ruler more useful for this new purpose — not of measuring lengths, but of helping us work out simple additions and subtractions.

In the first place, a 12-inch rule is inconvenient. It only exists because 12 inches make 1 foot, but we are not interested in that just now. We are interested in arithmetic and in arithmetic we base everything on the number 10.

Suppose you start with nothing (zero) and begin to count: 0, 1, 2, 3, 4, 5, 6, 7, 8, 9. Counting the zero, there are ten different symbols (or *digits*) for the first ten numbers. The next number, 10, makes use of two digits, with a 1 on the left. The 1 on the left is kept there in the next few numbers while the digit on the right goes through the same series over again: 10, 11, 12, 13, 14, 15, 16, 17, 18, 19. With a 2 on the left, we again go through the same series; then we do the same with a 3 on the left, with a 4 on the left and so on.*

* For a fairly detailed discussion of how our number system works, you might refer to another book in this series: *Realm of Numbers* (Houghton Mifflin, 1959).

If, therefore, we learn to work with the numbers from 0 to 9 to begin with, it will be easy to apply the knowledge to the higher numbers which, in a way, are merely repetitions of the first set. Consequently, a 10-inch rule will be sufficient.

Let's use our 10-inch rule to deal with the addition problem $2 + 3 = 5$. To begin with, we count off 2 inches from the left, but we don't really have to do any actual counting. If we count off 1 inch from the left, we end at the figure 1; if we count off 2, we end at 2; if we count off 3, we end at 3 and so on. Therefore, for the first number in the problem, $2 + 3 = 5$, we move straight to 2, without counting.

Starting at 2, we next have to count off 3, but why

Figure 2

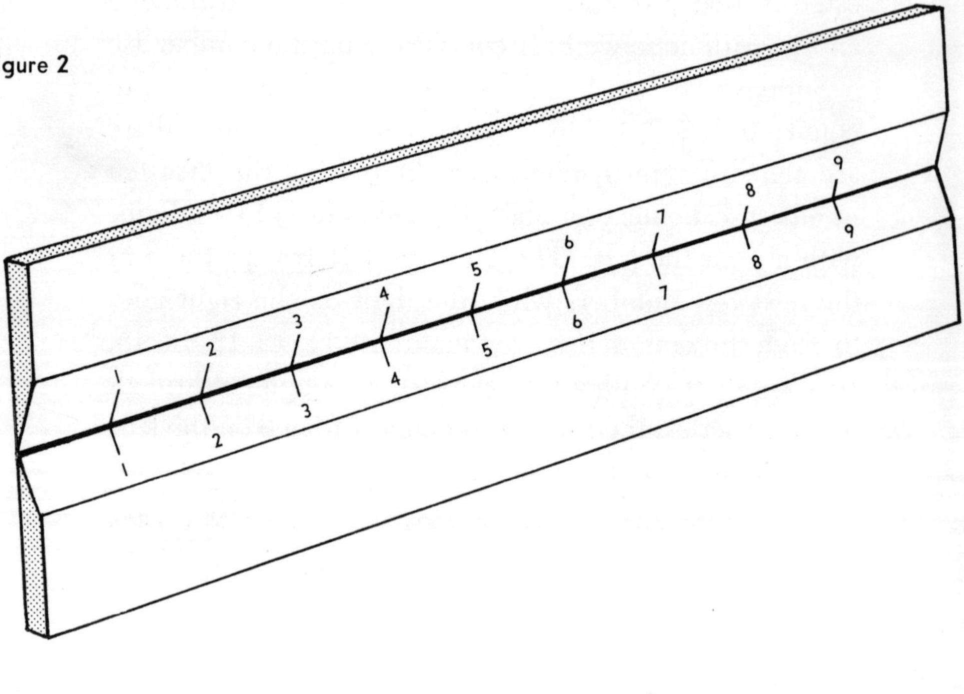

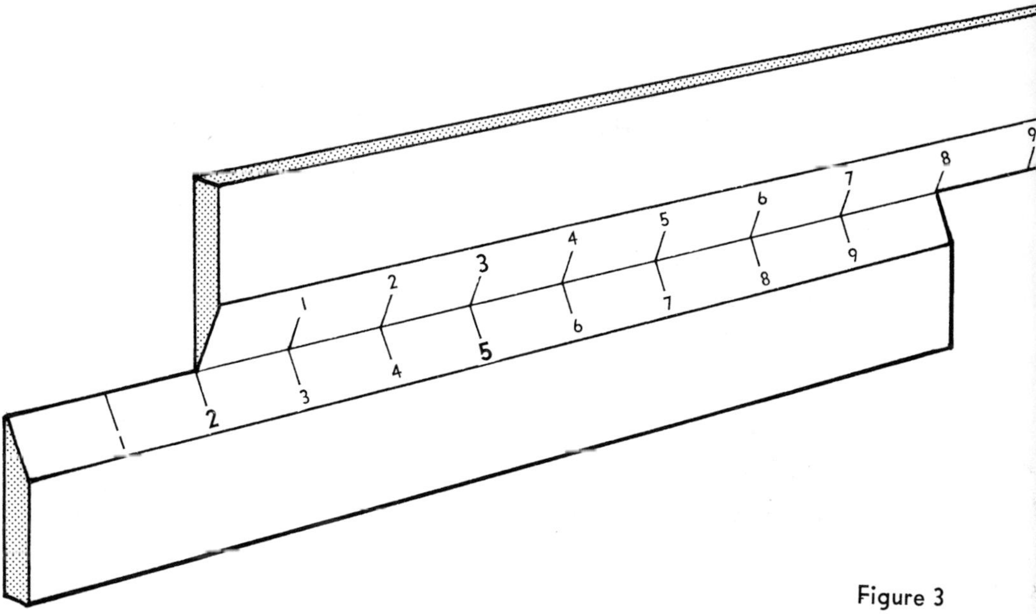

Figure 3

count it off? Since we are actually dealing with distances rather than with numbers, why not measure the 3 inches with another ruler?

Imagine a second ruler, with the inches marked off at the bottom, rather than at the top. Imagine the second ruler, edge to edge, on top of the first. The 1's will match on the two rulers, and so the 2's, the 3's, and so on (Figure 2).

Slide the top ruler to the right until its left end is over the 2 on the bottom ruler. Now, you see, the top ruler is measuring off the inches starting at the bottom 2 (Figure 3). If you look at the 3 on the top ruler, that marks a distance of 3 inches and immediately underneath it is the 5 of the bottom ruler. This is the method of showing that $2+3=5$ by measuring distances.*

If you keep the top ruler in the position shown in Figure 3, with its left end over the bottom 2, then you can see that $2+1=3$, because the top 1 is over the bottom 3, and, for similar reasons, $2+2=4$; and $2+4=6$. The single position of the ruler gives us the answers to a whole family of sums. It is easy to see how this system can be used to give answers to still other sums: $3+5=8$; $6+1=7$, and so on.

When two rulers are placed edge to edge, with one sliding along the other, the result is a slide rule. (You can also think of it as two sticks, one slipping along another. People who use a slide rule often, sometimes call it a "slipstick," but this is considered slang.)

A slide rule consisting merely of two ordinary rulers is anything but convenient, however. The ordinary ruler tapers to a narrow edge at the inch markings, in order to make it easier to draw a straight line and measure its length. Sometimes, the narrow edge even has a thin metal strip down its length to make it still easier to draw lines.

We don't use a slide rule, however, for drawing straight lines. We want to design it, instead, in such a way as to make it easy to slide one ruler against the

* It may seem silly to you to use so roundabout a method for finding out something as simple as $2 + 3 = 5$. Right now, however, we are only working out the technique. Things will remain very simple for quite a few pages, but before long we will be tackling more difficult problems — and solving them with just as little trouble as $2 + 3 = 5$ gives us now.

other. With thin edges, it is difficult to do this, and with metal strips it is practically impossible. It would be better to use two blunt rulers that are of even thickness all the way across, and that meet, therefore, at thick edges. (In fact, slide rules are usually a quarter of an inch thick.)

Even meeting at a thick edge has its problems. If the edges are quite smooth, the top ruler can easily slide sideways off the bottom one. For that reason, the bottom ruler is generally made with a groove down its length and the top ruler has a small tongue of wood that fits into the groove. Now the top ruler can slide back and forth easily, without any danger of slipping off sideways.

To be sure, the top ruler can still move upward easily. And if the slide rule happens to be turned upside down at any time, the top ruler, tongue and all, will fall out of the groove.

To prevent tipping, the top ruler can be made with a tongue of wood on top as well at on the bottom, and the upper tongue can be made to fit into a third ruler with a groove. What was the top ruler becomes a middle ruler. The next step is to bolt the uppermost ruler to the lowermost ruler, holding them firmly in place, while the now middle ruler can slide back and forth easily between them (Figure 4). The middle ruler cannot tip, and the slide rule cannot fall apart.

You might suppose that if you hold such a slide rule by one end and let it dangle, the middle ruler will simply slip out. However, the fit is usually made tight enough

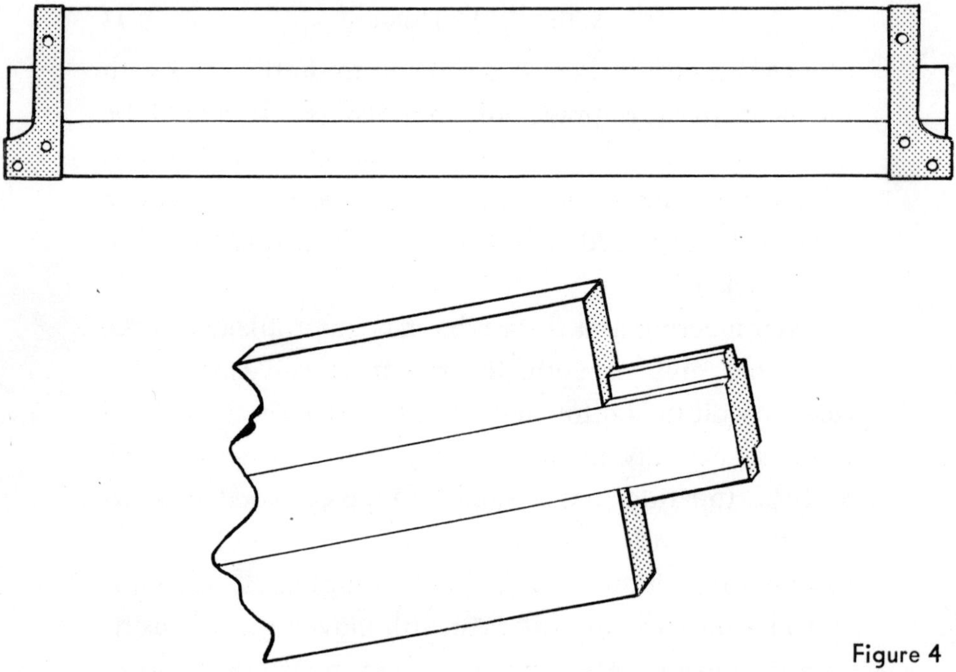

Figure 4

so that friction will keep it in place. To move that middle ruler, you actually have to push it.

Making the Rulers

Ordinarily, you could follow the directions I will give you for using a slide rule by following them on an actual instrument. That would serve as an additional guide to that of the illustrations that will be presented.

For a while, however, we are going to concern ourselves only with addition and subtraction, operations which are not conducted on the slide rules that are actually manufactured. (This is for the very good reason that addition and subtraction are too simple to require a slide rule.)

In fact, in describing the use of rulers in performing addition and subtraction, I hesitate to call the device a slide rule lest you confuse it with the slide rules actually manufactured for use in other operations. Let us therefore call the addition-and-subtraction device an *addition rule* and use only that term for the purpose.

In order to follow the workings of the addition rule as I describe them, you may have to rely very heavily on the illustrations. You cannot be guided by the use of an actual addition rule, unless you choose to make one for yourself. Fortunately, it is not difficult to make an addition rule.

To do this, get a sheet of flexible cardboard, of the type used in the manufacture of folders, and cut out two pieces (A and B) according to the measurements shown in Figure 5.

Fold Piece A along the lines *a* and *b*. The end pieces, which will overlap, are pasted together with mending tape, so that we now have something that looks like a flattened cylinder with a long rectangular hole in it.

Piece B will fit into Piece A snugly and can be moved back and forth within it. Piece A will now serve as the top and bottom ruler, "bolted" together with tape, while Piece B will be the middle ruler. Where Piece A and the lower portion of Piece B meet mark off inches from 0 to 10 on both as shown in Figure 6.

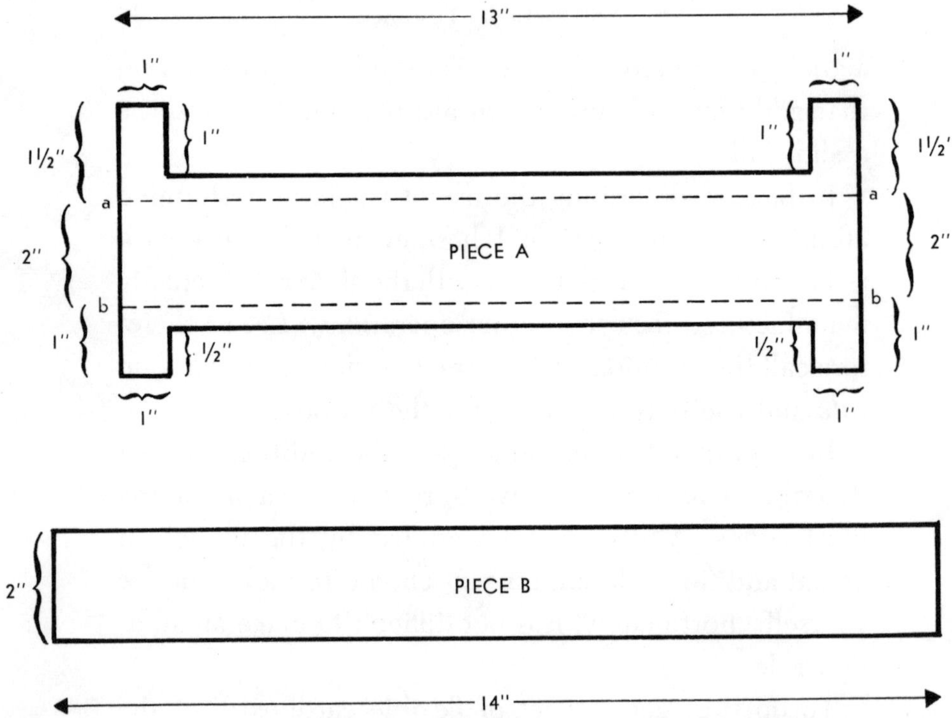

Figure 5

If you cannot find a piece of cardboard 14 inches long, you can make a smaller addition rule by working

Figure 6

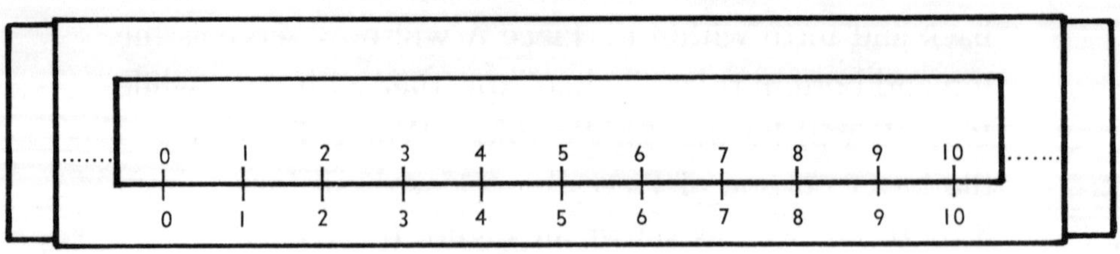

with measurements half the size of those indicated in Figure 5 and then marking off the 0 to 10 divisions in half-inch units.

The result, in either case, will be an addition rule which, however crude, will give you the feel of the instrument.

2

The Addition Rule

Beyond Ten

We have reached the point now where we should use proper terms, and no longer speak of top rulers, bottom rulers, and middle rulers. The part of the addition rule which slides back and forth is the *slide.* The rest of it is the *body.*

We can identify the range of numbers on the slide as the "S-numbers," and the range of numbers on the body as the "B-numbers." If I speak of S-7 and B-5, I mean the 7 marked on the slide and the 5 marked on the body.

What was the left end of an ordinary ruler — an end not marked by a number, as you can see if you look at one — is now represented by the number 0 on the addition rule.

Instead of saying, then, "Move the middle ruler to the right until its left end is over the 5 on the bottom ruler," I will say, much more simply, "Place S-0 over B-5."

In order to add 5 and 2 on the addition rule, the directions would be: Move S-0 over B-5, and under-

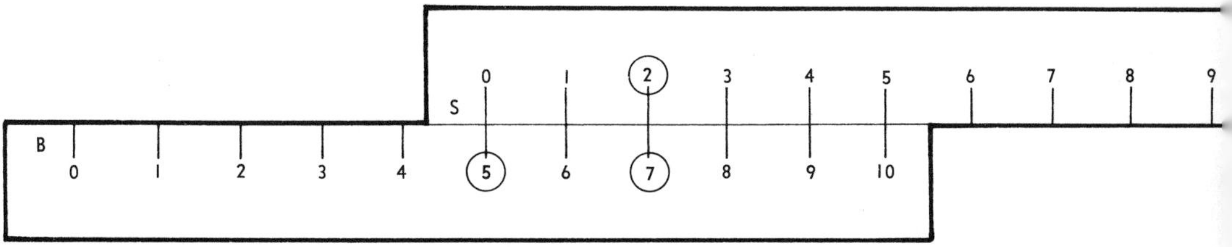

5+2=7

Figure 7

neath S-2, you will find B-7 (Figure 7), indicating that 5 + 2 = 7.

By now, though, you have undoubtedly realized that our addition rule will give us the answer to only the very simplest addition problems. Suppose we wanted the answer to 7 + 5, a problem which is still quite simple and which we know has the answer 12. Yet, at first glance, our addition rule can't help us.

If we place S-0 over B-7, then we would expect to find the answer under S-5. The trouble is, however, that S-5 is not making contact with the B-numbers at all (Figure 8). Now what?

One solution is to make the body of the addition rule longer, 20 units long, rather than 10. That would mean

Figure 8

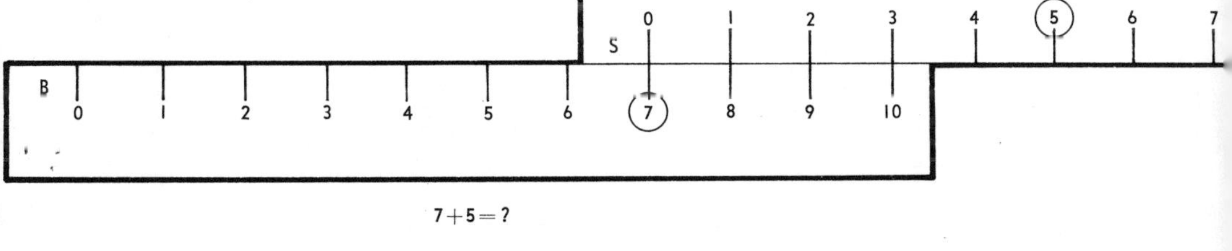

7+5=?

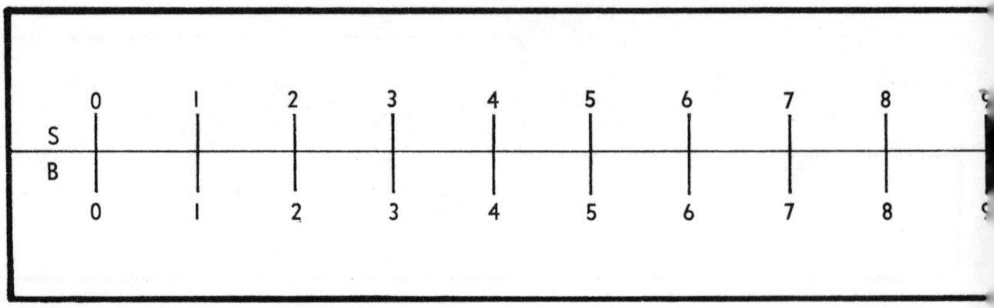

Figure 9

we can mark off numbers from 0 to 20. Let us remember, however, that the numbers 10 and beyond are repetitions, at least in their right-hand digits, of the numbers under 10. We can make this plainer if, in the numbers from 10 to 20, we place the digit to the left in parentheses, so that we can concentrate on the digit to the right (Figure 9).

Using this extended addition rule, we can add 7 and 5 by placing S-0 over B-7 and finding S-5 over B-(1)2. This shows us (Figure 10) that $7+5=12$.

The trouble is that the longer an addition rule is, the less convenient it is. No one would want to use one in which the bottom half of the body stuck out like an elephant's tusk. Is there any other way out?

To find one, look at Figure 10 again. Notice that S-0 is over B-7 and that S-(1)0 is over B-(1)7. Ignore the

Figure 10

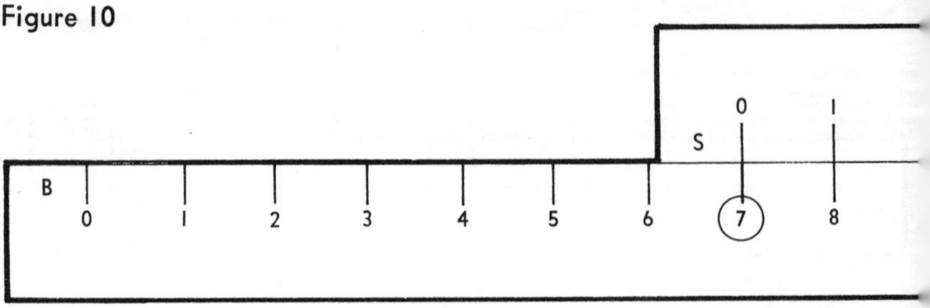

7+5=12

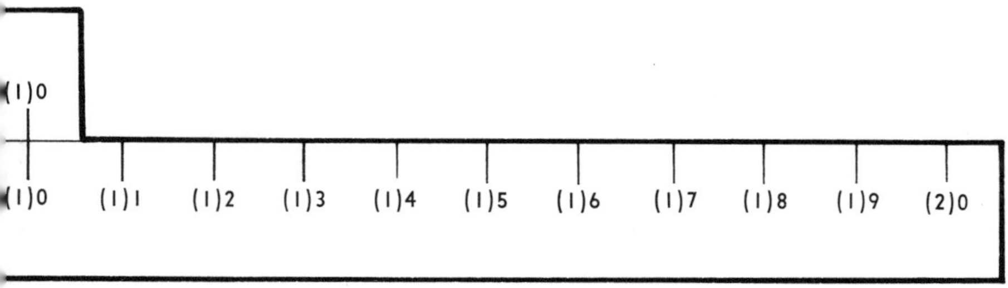

parentheses and you'll see that the situation has repeated itself. What's more, this would happen every time. If S-0 were over B-5, then S-(1)0 would be over B-(1)5; if S-0 were over B-2, then S-(1)0 would be over B-(1)2, and so on.

There's no puzzle about why this should be. In moving from S-0 to S-(1)0, we are adding 10 and we know that $2+10=12$, $5+10=15$, $7+10=17$ and so on. The right-hand digit of 10 is 0, and when we add 10 to any number, the right-hand digit of that number (to which we are adding zero) naturally remains unchanged.

In that case, it doesn't really matter whether we place S-0 or S-(1)0 over a B-number. In either case, the right-hand digit of the answer will be the same, and we must only remember to adjust the left digit. If we use S-0,

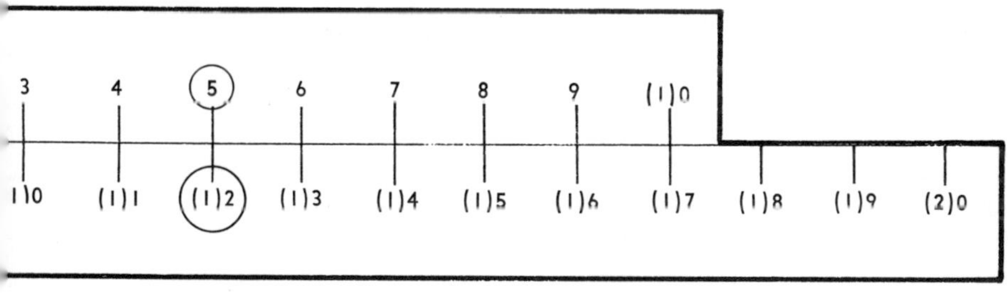

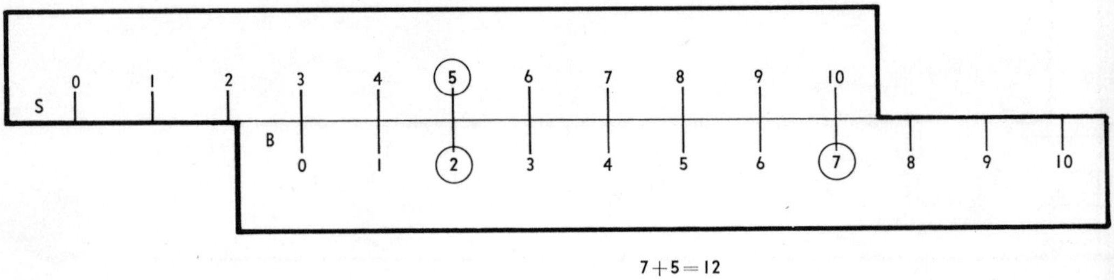

Figure 11

the left digit of the answer is 0 and is omitted. If we use S-10, the left digit of the answer is 1.

For instance, let's go back to our ordinary 10-unit addition rule and consider $7 + 5$ again. This time, instead of putting S-0 over B-7, we place S-10 over B-7 (Figure 11).

If you compare Figures 10 and 11, you will see how they duplicate each other. In Figure 10, on the 20-unit addition rule, S-4 is over B-(1)1, S-6 is over B-(1)3, and S-8 is over B-(1)5, while in Figure 11, on the 10-unit addition rule, S-4 is over B-1, S-6 is over B-3 and S-8 is over B-5.

On the 10-unit addition rule, if we place S-10 over some B-number, we need only add the left-hand digit 1 to our answer to get what we would have obtained with the much less convenient 20-unit addition rule. In Figure 11, with S-10 over B-7, we find that S-5 is over B-2. We add the left-hand 1 and we know that $7 + 5 = 12$.

We can also tell by Figure 11 that $7 + 4 = 11$, $7 + 7 = 14$, $7 + 9 = 16$.

Thus, by making use of both S-0 and S-10, we can

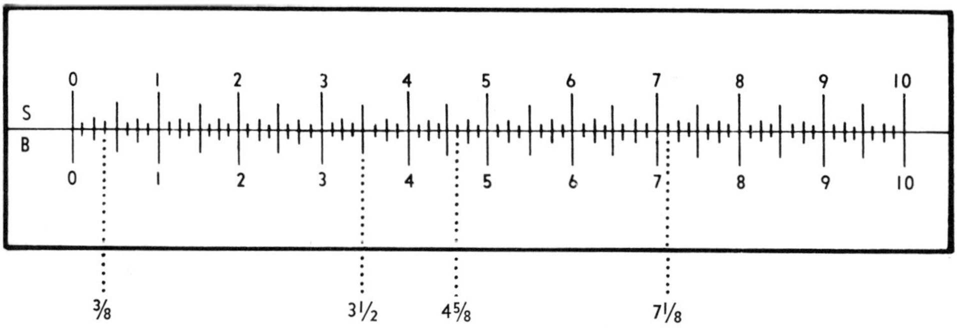

Figure 12

carry our additions on the 10-unit addition rule up to a sum of 20. And yet 20 isn't the highest possible sum there is either. What if we wanted to get the answer to $17 + 14$?

In order to tackle additions of this sort, let's take a detour in what seems the opposite direction — numbers smaller than 1 rather than larger than 20.

Fractions

So far, our addition rule only has digits on it, and we have been dealing only with a few small whole numbers. But you know that there are fractions — numbers that are smaller than one, or that are intermediate in value between neighboring whole numbers. In fact, if you look at an ordinary ruler again, you will find fractions there, for the inches are divided into halves, quarters, and eighths, and often sixteenths, too.

Suppose we divide each unit distance on our addition rule into halves, quarters, and eighths (Figure 12). You've undoubtedly had practice reading a ruler, so

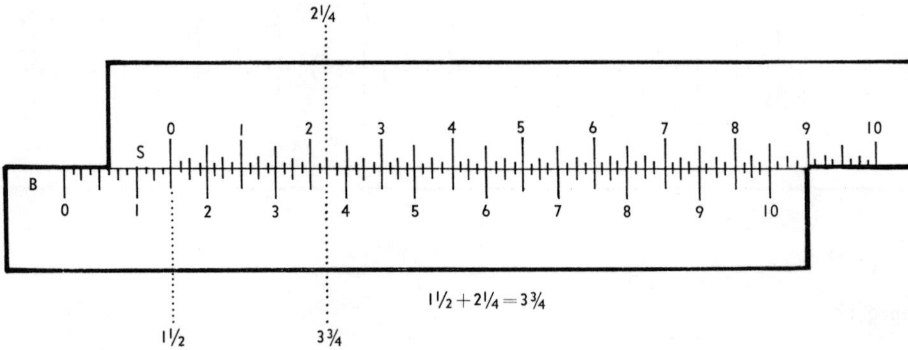

Figure 13

that you will probably have no difficulty in reading a particular mark in Figure 12 as 3½, or 4⅝, or 7⅛. And, of course, you will find the simple fraction ⅜ between 0 and 1.

Fractions work as well as whole numbers do on the addition rule. Why shouldn't they? If we add 1½ inches and 2¼ inches, we would get 3¾ inches, and that tells us 1½ + 2¼ = 3¾. Try it on the addition rule. Place S-0 over B-1½ and S-2¼ is found to be over B-3¾ (Figure 13).

By proper manipulation, you can find in the same way that 2⅜ + 4⅞ = 7¼; that 5¼ + 1⅞ = 7⅛, and so on. (The addition rule should impress you a little more now. It adds fractions as easily as it adds digits, whereas with pen and paper, it is considerably harder to add fractions than digits.)

For addition of fractions yielding sums over 10, we use S-10. To add 5⅜ and 7¼, for instance, we place S-10 over B-5⅜ and find S-7¼ over B-2⅝ (Figure 14). We

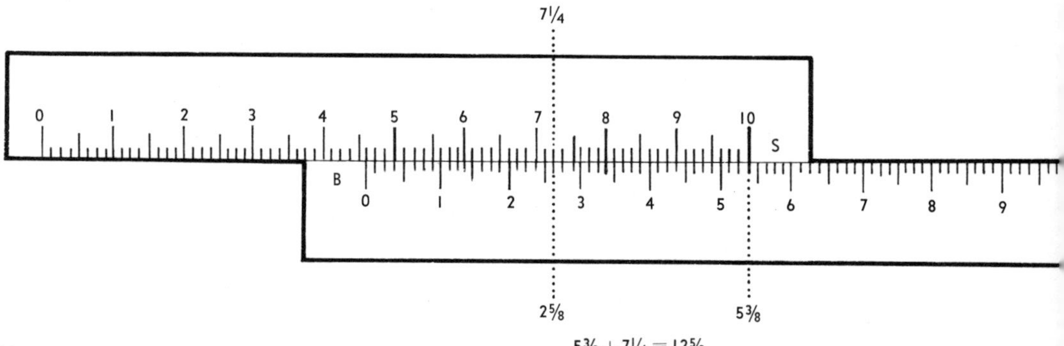

Figure 14

insert that left-hand digit 1, and find that $5\frac{3}{8} + 7\frac{1}{4} = 12\frac{5}{8}$.

There is no question that this system of adding fractions is very handy, but we can do still better. Remember that we changed the 12-inch ruler to a 10-inch ruler because in our number system the right-hand digits repeat after intervals of 10 (see page 7). This makes 10 a particularly useful number in calculations of all sorts. Following this system of emphasizing 10, we can divide each unit on the addition rule into tenths, instead of into halves, quarters, and eighths.

The result is shown in Figure 15. The $\frac{5}{10}$ mark (which

Figure 15

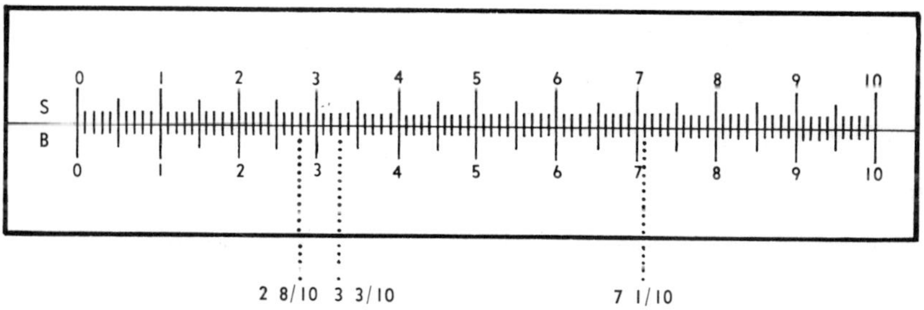

is equal to ½) is made longer than the others so that it stands out. You may be less accustomed to this system of subdividing units, but you should have no trouble picking out 3³⁄₁₀, or 7¹⁄₁₀, or 2⁸⁄₁₀.

It may seem at first that we have simply changed one set of fractions for another and you may wonder what we have gained. But tenths, you see, are special fractions because of our 10-based number system. With tenths, we can use decimal fractions. Thus 1¹⁄₁₀ can be written 1.1; 3⁶⁄₁₀ can be written 3.6; 9⁹⁄₁₀ can be written 9.9, and so on. By using tenths, then, we have switched our addition rule to the decimal system.

We can use the addition rule as easily for decimals as for ordinary fractions or for simple digits. If we want the sum of 1.7 and 1.4, we place S-0 over B-1.7 and find S-1.4 over B-3.1 (Figure 16). We can tell, then, that $1.7 + 1.4 = 3.1$. It is just as easy to discover that $2.3 + 7.1 = 9.4$, or that $5.8 + 1.6 = 7.4$.

Figure 16

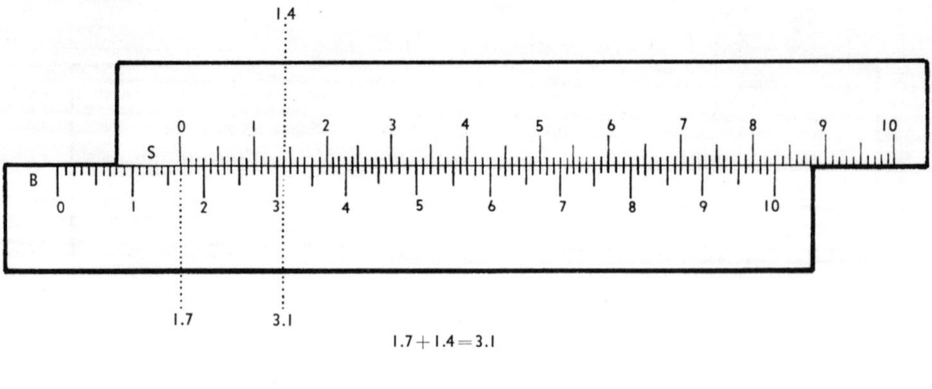

$1.7 + 1.4 = 3.1$

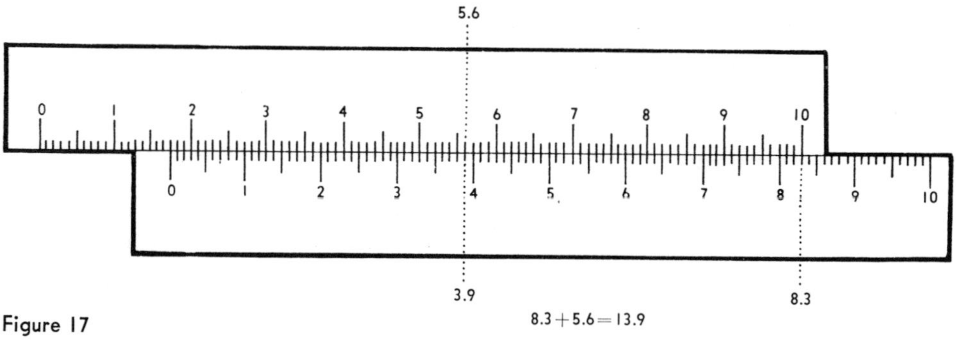

Figure 17

For sums over 10, we use S-10. If we place S-10 over B-8.3 and find S-5.6 over B-3.9 (Figure 17), we need only add the left-hand digit 1 to conclude that $8.3 + 5.6 = 13.9$.

Moving the Decimal Point

We are now ready to see a great advantage in the use of decimals over ordinary fractions. The decimals make it possible for us to handle additions that reach sums well over 20.

We have already determined by means of the addition rule that $1.7 + 1.4 = 3.1$ (Figure 16). Actually, though, we have discovered more than this. Changing the position of the decimal point doesn't affect the actual digits in the sum, provided the decimal point is moved in the same way in each number involved in the addition. If we moved each decimal point one place to the left, we would have $0.17 + 0.14 = 0.31$; and if we moved each one place to the right, we would have $17 + 14 = 31$. (This answers the question I asked on page 21 as to how the addition rule could be used to add 17 and 14.)

Suppose we want to add 83 and 56. We have already determined (Figure 17) that $8.3 + 5.6 = 13.9$. We can tell at once, therefore, that $83 + 56 = 139$.

You see, then, that the addition rule, subdivided in decimal fashion, can be used to add, with one shift of the slide, any two-digit number to any other two-digit number to give sums up to 200. This is better than merely reaching 20, but 200 isn't the highest number there is, either.

Suppose we divide each tenth on the scale into ten still smaller subdivisions. We would have tenths of tenths, or hundredths. We could then mark off numbers like 5.23 and 1.81 and we could find that $5.23 + 1.81 = 7.04$. By shifting decimal points mentally we would find that $52.3 + 18.1 = 70.4$ and that $523 + 181 = 704$. We could now add any three-digit number to any other three-digit number and get sums up to 2000.

But there's a catch, and that is that there's a limit to how finely you can divide your units. If you try to divide the space between the units into hundreds, then you will have tiny markings spaced a hundredth of an inch apart, and these would be far too difficult to read without close peering.

Actually, the smallest divisions into which an inch-wide unit can be divided conveniently are 0.02 inch units. Each tenth would have five subdivisions (Figure 18), so that you could read a number like 8.12 or 8.16.

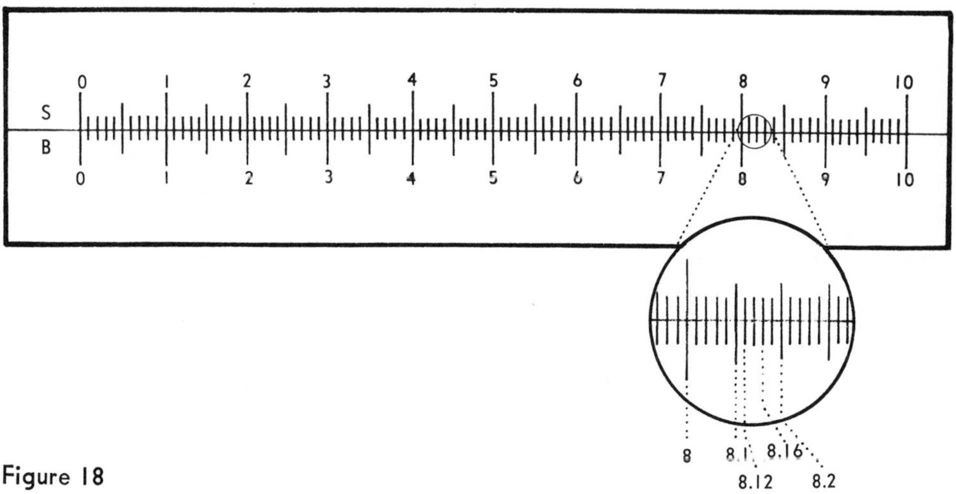

Figure 18

Using such an addition rule, you can find, without too much difficulty, that $0.24 + 8.12 = 8.36$ or that $8.12 + 6.72 = 14.84$. This will tell you that $2.4 + 81.2 = 83.6$ and that $81.2 + 67.2 = 148.4$. It will also tell you that $24 + 812 = 836$ and that $812 + 672 = 1484$.

It may seem to you that now we can only handle numbers with right-hand digits that are even, for the smallest divisions shown in Figure 18 represent 0.02, 0.04, 0.06 and 0.08. There is no division equivalent to 0.03, for instance, so that a number such as 7.33 could not be read directly.

And yet why not? You have no trouble finding 7.32 and 7.34 on the addition-rule as we now have it. Between them is an empty space with no markings and we can imagine a fine line splitting that empty space in two.

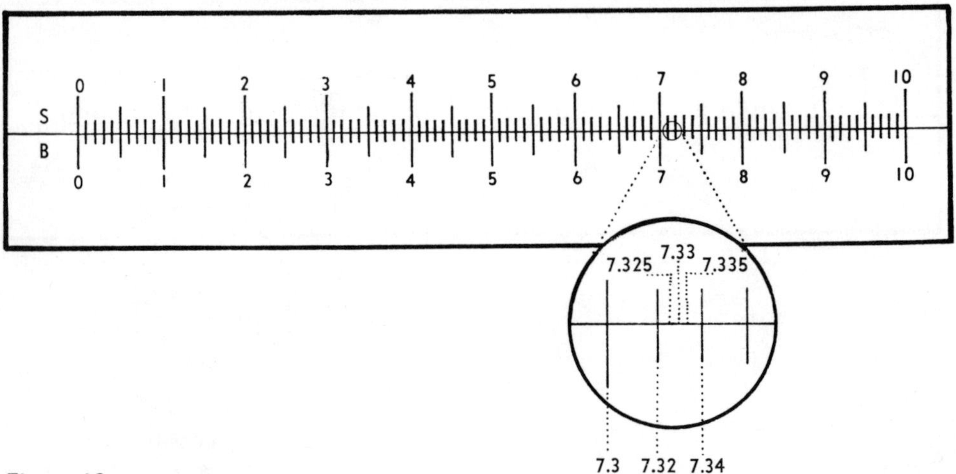

Figure 19

This imaginary line (shown in Figure 19 as a dotted line) would be 7.33. You might even imagine lines drawn closer to 7.32 than to 7.34, or vice versa and representing 7.325 or 7.335. These lines, which wouldn't be shown directly on even the most carefully manufactured 10-inch addition rule (if any were manufactured at all) can nevertheless be *estimated.*

To be sure, no matter what we do, there are limits to how finely we can read subdivisions. Our estimates are bound to be uncertain and we can't push them too far. In addition, the most carefully made markings may be very slightly off position and even if they aren't, the markings have a perceptible thickness (they must have) and ideally they should have no thickness at all.

We are, in short, condemned to inexactness. To see why that is, let's go back for a moment to ordinary fractions.

Inexactness

The decimal addition rule can handle ordinary fractions, too, provided those ordinary fractions are first converted to decimals. We can add 5½ and 2¾ without trouble if we convert the numbers to 5.50 and 2.75. Using the addition rule, we find that 5.50 + 2.75 = 8.25. If we wish, we can change 8.25 back to 8¼ and then we have determined that 5½ + 2¾ = 8¼.

The conversion of fractions to decimals, however, can present the addition rule with more than it can really handle. Suppose, for instance you wanted to add 3⅛ and 5⅓.

The decimal equivalent of 3⅛ is 3.125, and you have to estimate the position of 3.125 as an imaginary marking just one quarter of the way from 3.12 to 3.14.

As for 5⅓, that, in decimals, is 5.3333333 . . . and so on for any number of 3's you care to write. How do you locate that on the addition rule? You have the markings for 5.32 and 5.34, and the imaginary line halfway between is 5.33. Well, one-third of the way from 5.33 to 5.34 is the number you want.

Very well, then, you place S-0 on the B-3.125 and look under the S-5.333333 . . . and find yourself at a position just short of B 8.46. The position is somewhere between that mark and the imaginary mark that represents 8.45. You might estimate that it is three-quarters of the way over from 8.45 to 8.46 and therefore decide that the answer ought to be, well, 8.458.

But you have been estimating in three places now. You've estimated for 3.125, and for 5.33333, and now for 8.458. What is the real answer? If you work out the sum of $3\frac{1}{8}$ and $5\frac{1}{3}$ with pen and paper, you will find that the answer is $8\frac{11}{24}$ and if you change that into decimals it comes out to 8.458333333 . . .

The addition rule did not give you the right answer.

Look how close it came, though. It was off by only 0.0003333. You may feel hardhearted and say, "A miss is as good as a mile and a wrong answer is a wrong answer." But is it?

In the first place, in using the addition rule, you could find the almost-right answer in a moment, in a fraction of the time it would take you to get the exact answer by pen and paper.

To be sure, if you simply must have the exact answer and nothing else, the addition rule will have failed you, but often you can make do with less. It frequently happens in science, engineering, architecture, or in any field where numerical calculations are much used, that it is not necessary to get the absolutely exact answer. A very close answer will be fine. In that case, the addition rule with its close answer is what you need.

Then, too, even if you want the exact answer, the addition rule can still be useful as a check. Suppose you added $3\frac{1}{8}$ and $5\frac{1}{3}$, got the correct and exact answer of $8\frac{11}{24}$ and wanted to change that into decimals. You might perhaps get the answer 8.43842 through some arithmetical mistake involved in long division. (It is easy to make arithmetical mistakes in long division.)

If you then perform the same addition on the addition rule and find the answer is "just short of 8.46," you know something is wrong. If the correct answer were 8.43842, then the addition rule would say the answer was "just short of 8.44." The addition rule, if properly handled, cannot mistake 8.46 for 8.44, and you know you have made an arithmetical error in your pen-and-paper calculations and start checking it. So you see, the addition rule can be extremely useful even when it gives only approximate answers and not exact ones.

And, of course, it is important to realize that all through this book, I will be giving approximate answers. When I say that $3\frac{1}{8} + 5\frac{1}{3} = 8.458$, it is not intended to be an exact answer but an "addition-rule answer." It is sufficient that it is a very close answer.

Reversing the Process

And what about subtraction? Since, in arithmetic, subtraction is the opposite of addition, it would seem that any device that is capable of working out additions can work out subtractions, too, if it is run backward.

On page 6, for instance, I pointed out that you could solve $10 - 3$ on an ordinary ruler by counting off 10 inches and then moving back 3 inches to the figure 7. Can't that be done on the addition slide rule?

Your first impulse might be to place S-0 over B-10,

but what good would that do us? We can't count backward (that is, to the left) on the slide under such conditions since there is nothing to the left of S-0.

One way out would be to move some other part of the slide over the B-10. We might move S-7 over it and count back to S-4, or move S-5 over it and count back to S-2. Best of all, we might move S-3 over it and count back to S-0.

You can see that the last alternative is best, because it is automatic. The 3 is part of the problem, 10 − 3, and you are, in effect, subtracting the 3 from itself, leaving 0. Whenever you subtract a number from itself, you leave 0,

Figure 20

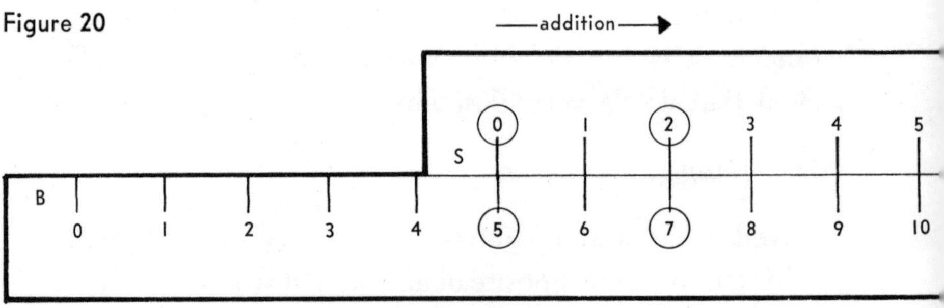

5+2=7

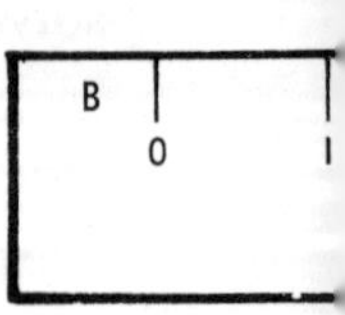

so that if you follow this procedure, you will always find the answer under S-0.

This fits in neatly with the process of addition, if we remember that subtraction is the reverse of addition. Consider two problems: $5 + 2$ and $5 - 2$. In the former, we place S-0 over B-5 and look *forward* to S-2, under which we find B-7. In the latter, we place S-2 over B-5 and look *backward* to S-0, under which we find B-3 (Figure 20). Thus we find that $5 + 2 = 7$ and $5 - 2 = 3$.

Here we run the danger of falling into a difficulty. In addition, it doesn't matter in which order we add our numbers. Faced with the problem of $5 + 2$, it doesn't

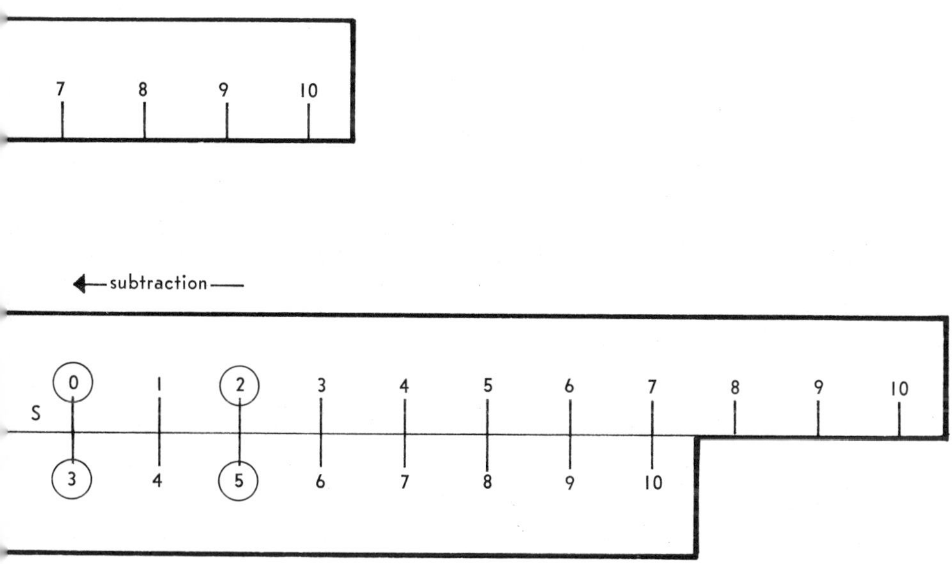

5 — 2 = 3

matter whether we treat it as $5 + 2$ or as $2 + 5$. The answer will be 7 in either case. This lends the addition rule a certain flexibility. If you place S-0 over B-5, the answer will appear under S-2. If you place S-0 over B-2, the answer will appear under S-5. In either case, you will find B-7 as the answer.

In subtraction, matters are less flexible. The answer to $5 - 2$ is not the same as the answer to $2 - 5$. In the former case, the answer is 3, in the latter it is -3.* We must, therefore, be careful about the manner in which we manipulate the addition rule if we are to solve subtraction problems correctly.

The best system I know of is to remember that the B-numbers are immovable while the S-numbers can be pushed back and forth and can therefore be viewed as changeable. In the problem $5 - 2$, the first number, 5 (the *minuend*), is the number you start with; a number which, if untouched, will remain 5 forever. It, therefore, belongs among the unchanging B-numbers. The second number, 2 (the *subtrahend*), introduces change; its presence alters the value of 5. The subtrahend, therefore, belongs among the changeable S-numbers.

* It is possible to design an addition rule to deal with negative numbers. To do this we would have the B numbers and S numbers extend leftward from 0 to give a series of ten negative numbers. You can then have readings of B-(-5) or S-(-3), for instance. However, such a negative extension will have no application to the real slide rule and its uses, which I will be dealing with shortly. We will, therefore, pay no attention to negative numbers.

In the general subtraction problem, $a - b = c$, then, you place S-b over B-a and then find B-c under S-*0*.

Following this system, you will see that to solve 5.84 − 2.28, you place S-2.28 over B-5.84 (never vice versa), and under S-0, you will find B-3.56. You know, therefore, that 5.84 − 2.28 = 3.56 or that (just as surely) 584 − 228 = 356.

There's another way of handling subtraction. Suppose the numbers on the slide ran in reverse, with the 0 on the extreme right, and then read leftward, 1, 2, 3, and so on, all the way to the 10 on the extreme left. This would be an *inverted scale,* in place of the ordinary *direct scale.* If you place the 0 of such an inverted scale over a B-number, you could then find room to look leftward and perform a subtraction.

We can call such a range of inverted numbers on the slide, SI-numbers (for "slide, inverted"). If we wanted to solve 5 − 2 using SI-numbers, we would place SI-0 over B-5 and, looking under SI-2, find B-3. If we wanted to add with the SI-numbers, we could solve 5 + 2 by placing SI-2 over B-5 and finding B-7 under SI-0 (Figure 21).

If you compare Figures 20 and 21, you will see that whether you use S-numbers or SI-numbers, you move to the right (forward) in addition and to the left (backward) in subtraction. The difference lies in the use of the 0. In addition, you place the S-0 over the number to which you are adding, but not SI-0. In subtraction, you

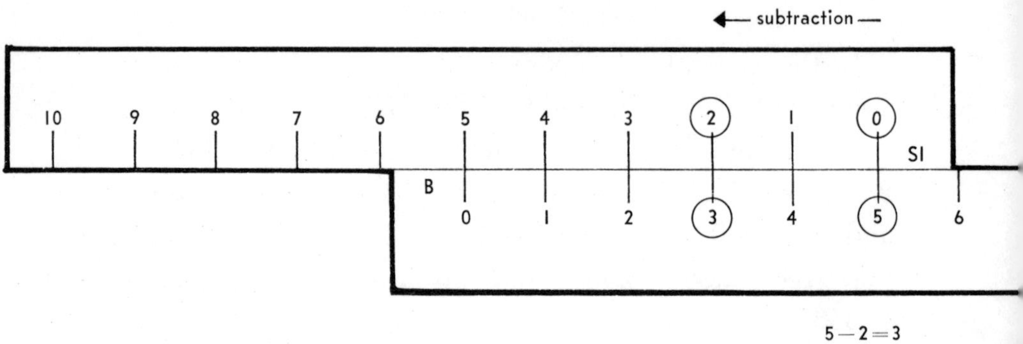

Figure 21

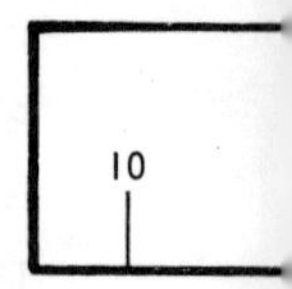

place SI-0 over the number from which you are subtracting, but not S-0. If, then, you want to make use of 0 in this fashion for both addition and subtraction, you must use the S-numbers for the first and the SI-numbers for the second.

This does not mean we must have an addition rule and a subtraction rule, as two separate devices. There is room for both sets of numbers, the S and the SI, on the addition rule we have been using.

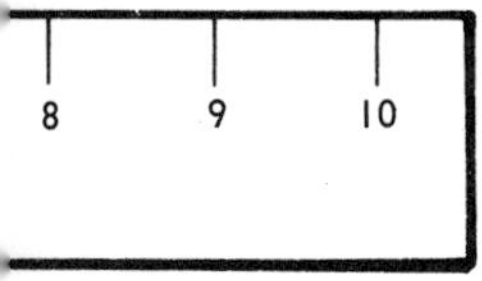

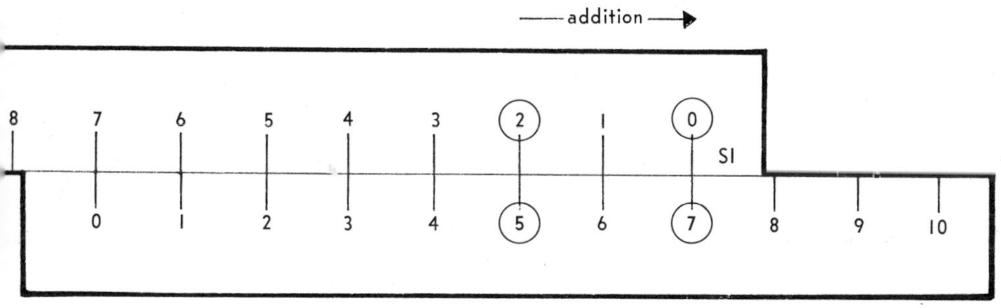

5 + 2 = 7

After all, there are two parts to the body of the addition rule, an upper part and a lower part, and so far we have been using only the lower part. There's nothing to prevent us from using the upper part, too.

Let's place a set of numbers on the upper part of the body, exactly like that on the lower part of the body. We can refer to the upper set as the UB-numbers ("upper body") to distinguish it from the ordinary B-numbers we have been using till now.

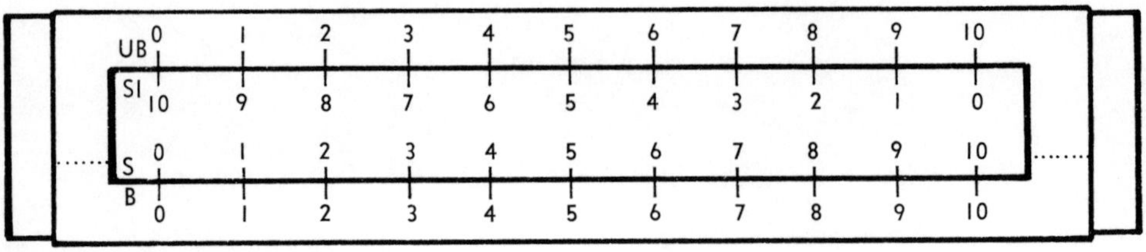

Figure 22

Along the upper edge of the slide, adjacent to the UB-numbers, are the SI-numbers. Thus we have two pairs of scales on the same addition rule, one for addition with S-

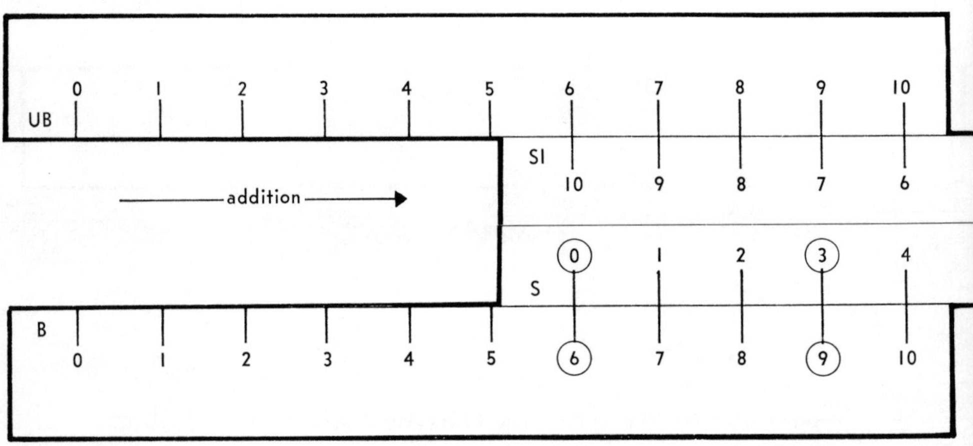

6+3=9

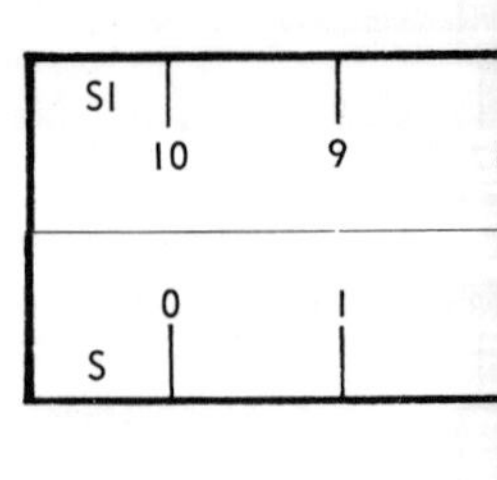

Figure 23

numbers and one for subtraction with SI-numbers (Figure 22.)

Let's use our double addition rule to work out $6 + 3$ and $6 - 3$ just to see how it works.

For addition, place S-0 over B-6 and look under S-3, where you will find B-9, so that you see that $6 + 3 = 9$. For subtraction, place SI-0 under UB-6 and look over SI-3, where you will find UB-3, so that $6 - 3 = 3$ (Figure 23).

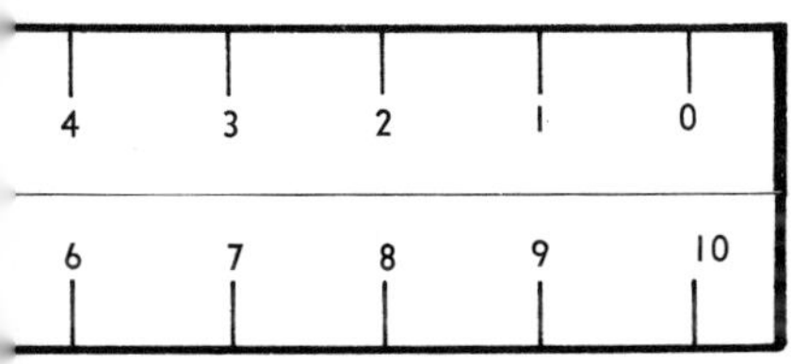

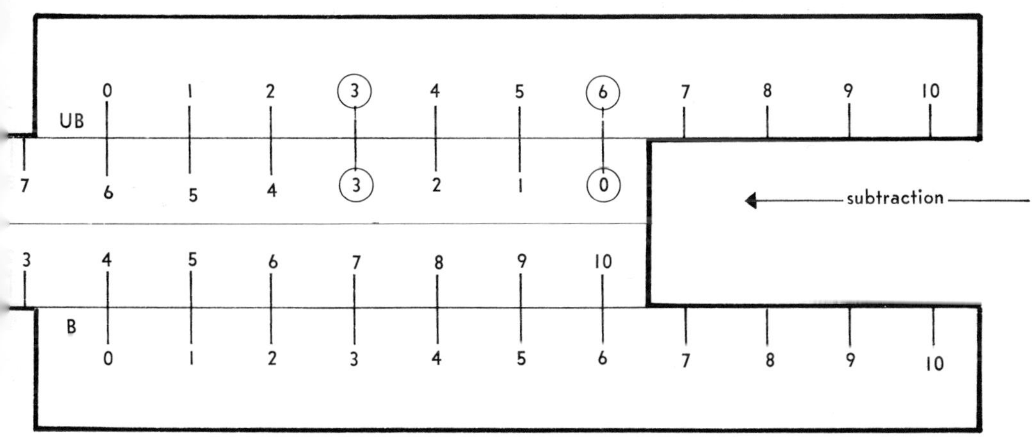

$6 - 3 = 3$

This brings to an end what I have to say about the addition rule. There is considerably more that could be written, but I now have all I need for the main business of this book.

As I said in the first chapter, addition rules are not manufactured, because addition and subtraction are such easy operations with pen and paper (except for occasional fractions) that no one bothers to take the trouble to manipulate an addition rule instead.

Please don't feel cheated at this. Don't feel you have learned various manipulations "for nothing." All the principles I have described in connection with the addition rule can also be used on actual slide rules which deal with operations more complicated than addition and subtraction.

On actual slide rules, however, you might have had trouble seeing why those principles worked, because the scales used are tricky ones. As it is, you learned the principles on the simplest possible scales, those made up of evenly spaced digits. I hope that it will now be easy to transfer the principles from the simple scales of the addition rule to the complex scales of the true slide rules.

3

Logarithms

Multiples of Two

ACTUALLY, the whole key to any slide rule, however fancy and complicated it may seem, is only addition and subtraction. The point is, however, that you must add and subtract special kinds of numbers. In this chapter, we will track down those special kinds.

Let us look, for instance, at the following set of numbers: 2, 4, 8, 16, 32, 64, 128, 256, 512 . . . You can continue the set as far as you wish, for, as you see, each number is just double the one before.

An odd thing about such a list of numbers is this: If you multiply any two of them, the product will be another number on the list. Thus, $2 \times 8 = 16$; $16 \times 32 = 512$; $4 \times 64 = 256$. Then, too, if we want to get into larger numbers, $512 \times 512 = 262{,}144$; and if you continue to work out the list of numbers by doubling each new value, you will find that the eighteenth number on the list is indeed 262,144.

This list (or "set") of numbers is, in other words, "closed to multiplication."

Let's look at the list in another way. Since we get each number by doubling the one before — that is, by multiplying the previous number by two — we can write the first number as 2, the second as 2×2, the third as $2 \times 2 \times 2$, the fourth as $2 \times 2 \times 2 \times 2$, and so on. The list becomes a set of multiples of 2.

Of course, it is clumsy to list the numbers as more and more 2's multiplied together, and it is natural to search for a simpler means of indicating this. The system commonly used is to write 2, standing by itself, as 2^1. The number 4, which is 2×2 (two 2's multiplied together), is 2^2. The number 8, which is $2 \times 2 \times 2$ (three 2's multiplied together), is 2^3. By this system, you would expect 2^5 to be $2 \times 2 \times 2 \times 2 \times 2$ (five 2's multiplied together) and if you work out the product you will find it to be 32. Therefore, $2^5 = 32$.

The small ^{5} (in the expression 2^5) is an *exponent* and 2^5 is an *exponential number*. Suppose, then, we try to express our set of numbers in exponential form. Instead of writing 2, 4, 8, 16, 32 . . . we can write 2^1, 2^2, 2^3, 2^4, 2^5 . . . The latter method is clearly the simpler and neater of the two.

Suppose we multiply these numbers in their exponential forms. Instead of writing $2 \times 8 = 16$, we can write $2^1 \times 2^3 = 2^4$. Instead of $16 \times 32 = 512$, we have $2^4 \times 2^5 = 2^9$. Instead of $512 \times 512 = 262{,}144$, we have $2^9 \times 2^9 = 2^{18}$.

If you will examine these multiplications, you will see that, in every case, the exponents have been added. In the case of $2^1 \times 2^3 = 2^4$, $1 + 3 = 4$. In the other two cases $4 + 5 = 9$ and $9 + 9 = 18$.

This is not mysterious. In the case of $2^4 \times 2^5$, you are multiplying a set of four 2's multiplied together ($2^4 = 2 \times 2 \times 2 \times 2$) by a set of five 2's multiplied together ($2^5 = 2 \times 2 \times 2 \times 2 \times 2$). If you multiply the first set by the second set, you end with a set of nine 2's multiplied together:

$$(2 \times 2 \times 2 \times 2) \times (2 \times 2 \times 2 \times 2 \times 2) = (2 \times 2 \times 2 \times 2 \times 2 \times 2 \times 2 \times 2 \times 2)$$

In multiplying sets of 2's in this fashion, you add the number of 2's in the various sets, and this is why exponents are added when exponential numbers are multiplied. We can make this general by saying that $2^x \times 2^y = 2^{(x + y)}$.

Next, we'll try something else. Take a number of our set of multiples of two and divide it by a smaller number of the set. The quotient is also a number of the set. Thus $256 \div 32 = 8$. Turn that into exponential numbers and it is $2^8 \div 2^5 = 2^3$. As you might guess, $8 - 5 = 3$.

This also is not surprising. If we divide 2^8 by 2^5, we are carrying through the following division:

$$\frac{2 \times 2 \times 2 \times 2 \times 2 \times 2 \times 2 \times 2}{2 \times 2 \times 2 \times 2 \times 2}$$

The five 2's in the denominator cancel five of the eight 2's in the numerator, leaving three 2's multiplied together as the answer. Therefore, $2^8 \div 2^5 = 2^3$ and, in general $2^x \div 2^y = 2^{(x-y)}$.*

We have thus discovered a method of converting certain multiplications into additions, and certain divisions into subtractions. Since it is the exponents which are added and subtracted in these cases, we are going to concentrate on these exponents from now on.

If we extract the exponent from an exponential number and set it down all by itself, it is customary to give it a new name and call it a *logarithm.* For instance, if we are considering the expression $2^3 = 8$, then 3, taken by itself, is the logarithm of 8. Of course, we don't want to forget the 2, which rests under the exponent as though it were a base. Consequently we speak of 3 as the logarithm of 8 to the base 2.

In the same way, the logarithm of 16 to the base 2 is 4, for $16 = 2^4$; and the logarithm of 32 to the base 2 is 5, for $2^5 = 32$.

We can abbreviate "logarithm to the base 2" as log_2. In that case, we can say: $log_2 8 = 3$, $log_2 16 = 4$, $log_2 32 = 5$, and so on.

Suppose we want to work it backward now. We have the logarithm to the base 2, and we want to write down the number it represents. If we have the logarithm 4 to

* This relationship between multiplication and division of exponential numbers — where exponents are added in the first case and subtracted in the second — is not surprising. Division is an operation which is the inverse of multiplication, just as subtraction is the inverse of addition. Keep this in mind for it will be handy later.

to the base 2, then the number it represents is 16, so that 16 is the *antilogarithm* of 4. In the same way if we have the logarithm 9 to the base 2 then its antilogarithm is 512, since $2^9 = 512$. The antilogarithm in these cases is to the base 2, and the expression can be abbreviated *antilog*$_2$. Thus, we can say that *antilog*$_2$ $4 = 16$, and *antilog*$_2$ $9 = 512$.

These new terms are confusing at first and will cease to be confusing only with practice. However, to be as clear as possible right now at the start, consider the expression $x^a = b$. In this general exponential expression, x is the base, a is the logarithm, and b is the antilogarithm. (It might help you to keep in mind the fact that logarithms are exponents, while antilogarithms are the "ordinary numbers" you use in everyday computations.)

It is easy to make a small table of logarithms and antilogarithms to the base 2, as follows:

log$_2$	*antilog*$_2$	*log*$_2$	*antilog*$_2$
1	2	11	2,048
2	4	12	4,096
3	8	13	8,192
4	16	14	16,384
5	32	15	32,768
6	64	16	65,536
7	128	17	131,072
8	256	18	262,144
9	512	19	524,288
10	1,024	20	1,048,576

You can continue the list as long as you like, but we have enough now to make our point. Remember that whenever we multiply two antilogarithms, we can achieve the same result by adding the corresponding logarithms. Suppose, for instance, we wanted to multiply 128 and 4096. These "ordinary numbers" are antilogarithms and may be found in the column headed *antilog*$_2$. The logarithms corresponding to them are 7 and 12 respectively. We add the logarithms and find that $7 + 12 = 19$, so that 19 is our logarithm sum. We find 19 in the *log*$_2$ column, and see that 524,288 is the corresponding antilogarithm. The logarithm sum is the antilogarithm product. We therefore conclude that $128 \times 4096 = 524{,}288$, a fact you can check by long multiplication.

Again, since $10 + 8 = 18$ (logarithms), we can see at once that $1024 \times 256 = 262{,}144$ (antilogarithms.)

We can try subtracting logarithms, too. Since $20 - 15 = 5$ (logarithms), we conclude that $1{,}048{,}576 \div 32{,}768 = 32$. Try this by long division and see if it is not right.

You will agree, I think, that it is much easier to add and subtract logarithms and make use of columns such as those given above, than it is to multiply and divide ordinary numbers (that is, antilogarithms).

Multiples of Ten

There are, however, flaws to this pretty picture. Let's begin with the biggest flaw of all. What I have described, so far, will only work for a few numbers which happen to be built up through the multiplication of 2's. You can deal with 32×64, but suppose you want to deal with 31×63. You are stuck. You can't produce either 31 or 63 by multiplying 2's.

One possible help might be to produce numbers which are multiples of other integers. For instance, you can't produce either 27 or 81 by multiplying 2's, but you can produce them by multiplying 3's. Thus, $27 = 3 \times 3 \times 3 = 3^3$ and $81 = 3 \times 3 \times 3 \times 3 = 3^4$. If you want the answer to 27×81, try it in exponential form. Since $3 + 4 = 7$, $3^3 + 3^4 = 3^7$. The number 3^7 represents the product of seven 3's multiplied together, and that works out to 2187. So you can say that $27 \times 81 = 2187$.

In short, you can prepare columns that list logarithms to the base 3 and their corresponding antilogarithms to the base 3. You can then solve problems which you can't solve by using logarithms to the base 2 in the fashion described above.

In the same way, we can work with logarithms to the base 5, or to the base 7, or to any base we may care to choose and in each case work with a new set of numbers.

However, this does not solve our problem. Every number can be found among the list of antilogarithms to one base or another, but what if you want to multiply a number from one list by a number from another list?

Consider, for instance, the problem $8 \times 9 = 72$. You can see that $8 = 2^3$ and $9 = 3^2$. This means that $log_2 8 = 3$, and $log_3 9 = 2$. These are logarithms to different bases. Does that matter? If they were logarithms to the same base, we would add them and achieve the same result that we would by multiplying the antilogarithms 8 and 9. Let's try adding them anyway. Since $2 + 3 = 5$, we decide that 5 is the logarithm of the product — but the logarithm to which base?

If 5 is a logarithm to the base 2, then the answer is 2^5 or 32. If it is a logarithm to the base 3, the answer is 3^5 or 243. But the actual answer, 72, is neither. In short, we simply can't add logarithms to one base and logarithms to another base, any more than we can add boys to cows.

If we are going to make logarithms useful, we must stick to one particular base and work out a way for finding logarithms to that base for all possible numbers. But if we are to do this, which particular base are we to use?

Perhaps at first, you might think that the most convenient possible base is 2, and in some ways it is. After all, doubling is so simple that it is much easier to prepare a list of multiples of 2 (2, 4, 8, 16, 32 . . .), than of 3's (3, 9, 27, 81, 243 . . .), or of 5's (5, 25, 125, 625, 3125 . . .) or of almost any other number. Furthermore, the numbers in the list of multiples of 2 are more closely spaced together than the numbers in the list of multiples of any other numbers, so there are fewer numbers not on the list for which we must find logarithms.

However, here as in many other cases, the fact that our number system is built upon the number 10 overrides everything else. Because our number system is built upon 10, it is easier to prepare a list of multiples of 10 than of any other number.

We begin with 10 itself. Then we have $10 \times 10 = 100$. After that is $10 \times 10 \times 10 = 1000$; $10 \times 10 \times 10 \times 10 -$ 10,000, and so on. Our list is 10, 100, 1000, 10,000 . . . each number possessing one more zero than the number before. There is nothing easier than adding one more zero for each number on a list.

We can express such a list in exponential form, too, as 10^1, 10^2, 10^3, 10^4 . . . The connection between the ordinary 10-multiple and its exponential form is a simple one. Since $10 = 10^1$, $100 = 10 \times 10 = 10^2$, $1000 = 10 \times 10 \times 10 = 10^3$, and so on, you can see that the exponent is equal to the number of zeros in the number itself. You can see without having to go into great feats of computation that $10{,}000{,}000 = 10^7$ and $100{,}000{,}000{,}000 = 10^{11}$.

The simplicity of this connection extends to logarithms to the base 10. Remember that the logarithm is the exponent. Therefore since $100 = 10^2$, $log_{10}\, 100 = 2$. Again, since $10{,}000{,}000 = 10^7$, $log_{10}\, 10{,}000{,}000 = 7$. Again, the logarithm to the base 10 is equal to the number of zeros in a number of this sort.

Such logarithms are so easy to work out that logarithms to the base 10 are used far more often than logarithms of any other kind.* Consequently, when people

* There is a set of logarithms to another base very frequently used in higher mathematics. Since this book does not involve higher mathematics, we will not need to consider this other set, sometimes referred to as "natural logarithms."

speak of "logarithms" without specifying the base, they are almost certain to mean logarithms to the base 10.

For the rest of the book, logarithms to the base 10 are the only ones I shall use and I shall refer to them merely as logarithms, with the simple abbreviation *log*.

Logarithms of this sort are so easily obtained that one doesn't even need a set of columns of logarithms and antilogarithms to carry through multiplication and division of multiples of ten. Suppose we consider $10{,}000 \times 100$. By counting zeros, we know at once that *log* $10{,}000 = 4$ while *log* $100 = 2$. Since $4 + 2 = 6$, we know the logarithm of the product to be 6. That means there are 6 zeros in the corresponding antilogarithm, so that $10{,}000 \times 100 = 1{,}000{,}000$.

Again, if we consider $10{,}000 \div 100$, we must subtract logarithms. Since $4 - 2 = 2$, the logarithm of the quotient is 2 and the corresponding antilogarithm is 100. Therefore $10{,}000 \div 100 = 100$.

Between the Multiples

Of course, we can't be completely enthusiastic about all this, for so far we can only make use of the 10-multiples, such as 100 and 1000, and these are far fewer than the 2-multiples. If you inspect the columns on page 45, you will see that there are twenty 2-multiples up to the neighborhood of a million. There are only six 10-multiples in that same range.

Furthermore, the 10-multiples are particularly easy to

handle even without logarithms. It is no problem at all to decide, without logarithms, that $100 \times 1000 = 100{,}000$. What we need is some easy method of carrying through a multiplication such as 72×263.

We must therefore extend the notion of logarithms to numbers that don't end with zeros.

Let's begin by considering the following: $10 \div 10 = 1$. We can express 10 in exponential form as 10^1 and then express the division as follows (remembering to subtract exponents): $10^1 \div 10^1 = 10^0$. If we solve a problem, correctly, in two different fashions and get an answer in two different forms, those forms ought to represent the same number. Here we divide 10 by 10 and get two answers: 1 and 10^0. It is reasonable then to suppose that $10^0 = 1$. If we bring the exponent down, it becomes a logarithm and we can say that $\log 1 = 0$.

We know now that $\log 1 = 0$ and $\log 10 = 1$. What about the logarithms of numbers between 1 and 10? It would seem that those logarithms ought to lie between 0 and 1 in value. They ought, in other words, to be fractions.

But what meaning could a fractional logarithm have? Suppose there were a number which had a logarithm equal to ½. What kind of number would fit such a logarithm?

Let's consider a number x, such that $\log x = \frac{1}{2}$. Remember that a logarithm is simply an exponent brought down from its position above. If $\log x = \frac{1}{2}$, then $x = 10^{1/2}$.

We are certainly entitled to wonder what a number like

$10^{1/2}$ can possibly mean. Suppose we consider the product of $10^{1/2} \times 10^{1/2}$. If we add exponents we can see that $10^{1/2} \times 10^{1/2} = 10^1 = 10$.

We can therefore say that $10^{1/2}$ is that number which, when multiplied by itself, gives 10 for an answer. Such a number is the "square root of 10" and is usually written $\sqrt{10}$.

It is not difficult to work out an approximate value of the square root of 10. Actually, it turns out to be an unending decimal expression, but it is possible to work it out to as many decimal places as are desired, and the more decimal places that are worked out, the closer the number is to the actual value of the square root of 10.

Worked out to six decimal places, the square root of 10 is 3.162120. It turns out that $3.162120 \times 3.162120 = 9.9990028944$, which is almost 10 as you see. In fact, it would be pretty useful to speak of the square root of 10 as 3.162. This number, multiplied by itself, yields the product 9.998244, which is still pretty close to 10.

Since 3.162×3.162 is just about equal to 10, we can say that 3.162 is just about equal to $10^{1/2}$. We can therefore say, with a good approximation to the truth, that *antilog* $\frac{1}{2} = 3.162$ and that *log* $3.162 = \frac{1}{2}$.

Next let's consider that $10^{1/3} \times 10^{1/3} \times 10^{1/3} = 10^1 = 10$. Therefore, $10^{1/3}$ is a number which yields 10 when multiplied by itself twice. We can, therefore, say that $10^{1/3}$ is equal to the "cube root of 10" or $\sqrt[3]{10}$. It is possible to work out the cube root of 10 and this turns out to be (to three decimal places) 2.154. Therefore we can say, as an approximation, that *antilog* $\frac{1}{3} = 2.154$ and that *log* $2.154 = \frac{1}{3}$.

By adding exponents, we can see that $10^{1/3} \times 10^{1/3} = 10^{2/3}$. We already know that $10^{1/3}$ is equal to approximately 2.154. If we multiply 2.154 by itself we get about 4.64. Therefore, $10^{2/3}$ is equal to about 4.64, which means we can say that (approximately, at least) *antilog* $\frac{2}{3} = 4.64$ and *log* $4.64 = \frac{2}{3}$.

You can, if you choose, express such fractional logarithms in decimal form; and, indeed, they usually are so expressed. Instead of saying *log* $3.162 = \frac{1}{2}$, you would say *log* $3.162 = 0.5$. If you worked out the fact that *log* $1.585 = \frac{1}{5}$, you could express it as *log* $1.585 = 0.2$, and so on.

The cases I've presented above are examples of ways in which simple fractional logarithms can be obtained. Those simple fractions have as their antilogarithms rather complicated decimals. Mathematicians, however, have worked out methods for obtaining the logarithm of any number at all, including logarithms for all the simple digits. Thus, the logarithm of 2 turns out to be about 0.301 (to three decimal places), so we can say *log* $2 = 0.301$.

Actually, the logarithm of a number, in almost every case, is an unending decimal, but mathematicians can work out as many places as they choose. They can then prepare *logarithm tables,* in which the logarithms of a

set of consecutive numbers are presented. Such logarithms are commonly presented to five decimal places.*
By using such a table, we can find that:

antilog	*log*
1	0.00000
2	0.30103
3	0.47712
4	0.60206
5	0.69897
6	0.77815
7	0.84510
8	0.90309
9	0.95424
10	1.00000

This is equivalent to saying that we can place each of the digits from 1 to 10 into exponential form. We can say that $3 = 10^{0.47712}$, $7 = 10^{0.84510}$ and so on.

The usual five-place logarithm table will, of course, also give the logarithms of decimal numbers. For instance, such a table will tell you that *log* 4.354 = 0.63889 and that *log* 2.189 = 0.34025. That means that $4.354 = 10^{0.63889}$ and that $2.189 = 10^{0.34025}$.

In multiplying exponential numbers, we add exponents — which is to say, we add logarithms. Suppose, for

* If this were a book on logarithms, I would include such a table and give complete instructions for its use. However, I am using logarithms only to explain the workings of the slide rule. I will ask you, therefore, to accept, more or less on faith, the values of the logarithms as I present them in the meanwhile.

instance, you wanted the answer to 4.354×2.189. Instead of multiplying the numbers themselves, you add the logarithms and find that $0.63889 + 0.34025 = 0.97914$. The sum 0.97914 is the logarithm of the product of 4.354×2.189. In the logarithm table, we find that 0.97914 is the logarithm of 9.531. Therefore, we conclude that $4.354 \times 2.189 = 9.531$.

This is not the exact answer. The manipulation of logarithms can't give you the exact answer unless the exact values of the logarithms are used, and logarithms are virtually never known exactly. However, five-place logarithms are accurate enough for most purposes. For instance, if you work out 4.354×2.189 with pencil and paper, you find the exact answer is 9.530906 and surely 9.531 is quite close.

In the same way, you can subtract logarithms instead of dividing numbers. Suppose you wanted the answer to $4.354 \div 2.189$. Taking the logarithms again, we find that $0.63889 - 0.34025 = 0.29864$. The logarithm-difference 0.29864 is the logarithm of the quotient of $4.354 \div 2.189$. The logarithm table tells us that 0.29864 is the logarithm of 1.989. Therefore, we conclude that $4.354 \div 2.189 = 1.989$. If you work out the problem by long division, you find the answer is actually about 1.98858, but again 1.989 is reasonably close.

With practice, one could learn to use logarithm tables so quickly and easily that there would be no question of working out complicated multiplications and divisions in the ordinary way — unless one simply had to have absolutely exact answers. Even then, the logarithmic answer would be a convenient check against the ever-present possibility of arithmetical error.

Changing the Characteristic

So far, we have been working with the logarithms of the numbers from 1 to 10. There is, however, a vast array of numbers greater than 10 and smaller than 1. What about those?

What might the logarithm of 52.38 be, for instance?

We can solve that problem by considering 52.38 as 5.238×10. The logarithm of 5.238 is, according to the logarithm table, 0.71917. The logarithm of 10 is, of course, 1. If we are multiplying two numbers to get a product, we can just as well add the logarithms of those two numbers to get the logarithm of the product. Since $5.238 \times 10 = 52.38$, and $0.71917 + 1 = 1.71917$, then 1.71917 is the logarithm of 52.38. We can express this more simply by saying $\log 52.38 = 1.71917$.

It is simple to find the logarithm of 523.8 now. We see at once that $5.238 \times 100 = 523.8$, and we know that the logarithm of 100 is 2. Therefore using logarithms, $0.71917 + 2 = 2.71917$ and $\log 523.8 = 2.71917$. In the same way, we can show without trouble that $\log 5238 = 3.71917$, that $\log 52{,}380 = 4.71917$, and so on.

Let's make a small table to make sure we see the point clearly:

antilog	*log*
5.238	0.71917
52.38	1.71917
523.8	2.71917
5,238	3.71917
52,380	4.71917

We can divide the logarithm into two parts: the number to the left of the decimal point, which is called the *characteristic*, and the number to the right of the decimal point, which is called the *mantissa*. If we say that *log* 52.38 – 1.71917, the characteristic of that logarithm is 1, the mantissa is 71917.

When two numbers differ only in the position of the decimal point, the mantissa of their logarithms is identical, as you see in the example given in the table above. The characteristic, on the other hand, changes with the shifting decimal point in a number.

The change in the characteristic presents no problem. By inspecting the table above you can see that in the case of those numbers, at least, the value of the characteristic is one less than the number of digits to the left of the decimal point in the antilogarithm; and this it turns out is true in all cases.

This rule is so simple that we need no logarithm table to determine characteristics. By simply looking at the number 35.62, we see that there are two digits to the left of the decimal point and that the characteristic is therefore 1. In the same way we know that 282,100,000, which has nine digits to the left of the decimal point, has a characteristic of 8.

All we need the logarithm table for, then, is to determine the mantissa, and it should, therefore, really be called a "mantissa table." Since the mantissa is the same for any particular combination of digits, regardless of the position of the decimal point, we can ignore the decimal point altogether in finding the mantissa.

Suppose we want the logarithm of 35.62. We look up the mantissa for the digit-combination 3562 and find it to be 55169. Since 35.62 has two digits to the left of the decimal point, the characteristic of its logarithm is 1; and therefore $log\ 35.62 = 1.55169$.

In the same way, to get the logarithm of 282,100,000, we look up the mantissa for the digit-combination 2821 and find it to be 45040. Since 282,100,000 has nine digits to the left of the decimal point, the characteristic of its logarithm is 8. Therefore, $log\ 282{,}100{,}000 = 8.45040$.

Suppose, then, we wanted to multiply 123.1 and 35.2. We know at a glance that the characteristic of the logarithms of these two numbers are 2 and 1 respectively. We find the mantissas of the two digit-combinations 1231 and 3520 in the log tables and find them to be, respectively, 09026 and 54654. Therefore $log\ 123.1 = 2.09026$ and $log\ 35.2 = 1.54654$. If we add these two logarithms, $2.09026 + 1.54654$, we get 3.63680, which is the logarithm of the product of 123.1×35.2.

The problem, now, is to find the antilogarithm of

3.63680. It is the mantissa that will give us the necessary digit-combination, and the logarithm table tells us that the digit-combination of the antilogarithm represented by the mantissa 63680 is 4333. The characteristic, 3, tells us that we must place the decimal point so that there are four digits to its left. The answer is, therefore, 4333. (The actual answer, obtained by multiplying 123.1 and 35.2 in full, is 4333.12.)

We can divide similarly. If we want to divide 123.1 by 35.2, we subtract the logarithm of the latter from that of the former. It turns out that $2.09026 - 1.54654 = 0.54372$. The digit-combination of the antilogarithm of the mantissa, 54372, is 3497. Since the characteristic is 0, the decimal point must be so placed as to have one digit to its left. We can therefore say that $123.1 \div 35.2 = 3.497$. (If we work out $123.1 \div 35.2$ by long division, the answer turns out to be about 3.49715.)

And what about numbers less than 1? Suppose, for instance, we are faced with determining the logarithm of 0.481. We use the same strategy that worked for us before. We will consider 0.481 as $4.81 \div 10$. The logarithm of 4.81 is 0.68215 and the logarithm of 10 is 1. In dividing numbers we must subtract logarithms to get the logarithm of the quotient. Therefore, if $4.81 \div 10 = 0.481$ and $0.68215 - 1 = -0.31785$, then -0.31785 is the logarithm of 0.481.

This presents us with a slight problem. The mantissa of −0.31785 is 31785, which is different from the mantissa of the logarithm of 4.81. Furthermore the characteristic is −0, whatever that means. Fewer problems are created if we leave the logarithm of 0.481 as 0.68215 − 1, without trying to carry through the subtraction. That leaves the mantissa the same as in the logarithm of 4.81 or 48.1 or any of that family. And we can call −1 the characteristic.

In the same way, we find that 0.0481 can be written as 4.81 ÷ 100, so that its logarithm is 0.68215 − 2, the logarithm of 0.00481 is 0.68215 − 3, and so on.

If we write positive numbers less than 1, with one zero to the left of the decimal point, we can then work out a simple rule for determining the characteristic of their logarithms. The characteristic is a negative number with its value equal to the number of consecutive zeros that begin the antilogarithm.

For instance, 0.235 begins with a single zero, and the characteristic of its logarithm is −1. The number 0.000442 begins with four consecutive zeros and the characteristic of its logarithm is −4.

It is only consecutive zeroes that count in this rule. The number 0.02004, has four zeros all together, but it starts with two consecutive zeros and its characteristic is −2.

We find the mantissa in the ordinary way. For 0.235, we look up the mantissa of the digit combination 235 and find it to be 37107. Since the characteristic is −1, *log* 0.235 = 0.37107 − 1.

Again, to find the logarithm of 0.000442, we look up the mantissa of the digit-combination 442. This turns out to be 64542, so that *log* 0.000442 = 0.64542 − 4. To get the logarithm of 0.02004, we look up the mantissa of 2004 and find it to be 30190, so that *log* 0.02004 = 0.30190 − 2.

We can make use of these logarithms with negative characteristics in quite the usual way. If we wish to multiply 235 by 0.000442, we need but add the logarithms: 2.37107 + (0.64542 − 4) and this comes out to 3.01649 − 4, or, simplifying it, 0.01649 − 1. The mantissa 01649 corresponds to the antilogarithm digit-combination 10387 and from the characteristic of −1, we know the decimal point should be placed so as to make the answer 0.10387.

If we were considering the problem 235 ÷ 0.000442, we would subtract the second logarithm from the first. This means 2.37107 − (0.64542 − 4). As you probably know, subtracting a negative number is the same as adding a positive one, so that by removing parentheses, we have: 2.37107 − 0.64542 + 4 = 5.72565. The mantissa 72565 gives us 5317 as an antilogarithmic digit-combination. From the characteristic 5, we know that the answer must have six digits to the left of the decimal place and must therefore be 531,700.

Let us summarize, then. For any positive number, we can work out the characteristic of the logarithm at a glance. The mantissa can be found in a logarithm table and will be the same for any particular digit-combination regardless of the position of the decimal point.*

* This is true, in our number system, only for logarithms to the base 10 — another reason for preferring the base 10 to any other in ordinary computations. Nor is this a coincidence. It is the inevitable consequence of the fact that our number system is built around 10 and multiples of 10.

4

Logarithms on Wood

The L-Scale

LET US stop now and consider the situation. In Chapter 2, I showed how we could add and subtract numbers, automatically, on an addition rule. In Chapter 3, I showed how we could multiply and divide numbers by adding and subtracting their logarithms.

Could we not combine these two processes? Could we not use an addition rule to add and subtract logarithms and thus carry through multiplication and division automatically?

Indeed we can. An addition rule used to add and subtract logarithms is the true slide rule, and it is this slide rule which this book is intended to explain.

The simplest slide rule we can imagine is, actually, identical with an addition rule. Suppose, for instance, we are faced with the necessity of performing a particular multiplication. We use a logarithm table to look up the logarithms of the numbers being multiplied. We then add those logarithms on the slide rule, thus obtaining the logarithm of the product. We return to the logarithm table to get the antilogarithm of that final logarithm, and thus obtain the product of the multiplication.

This is clearly not very good. The time-consuming part of working with logarithms is the finding of logarithms and antilogarithms in the table. Adding the logarithms once they are found is very simple. We are thus making the slide rule do the easy part and neglect the tedious part.

But what if we could make the slide rule itself into a kind of logarithm table? It might then be unnecessary to look up logarithms and antilogarithms at all. We could make the slide rule look them up, so to speak, and, having found them, add or subtract them, and then give us the antilogarithm as the answer.

Can this be done? Or is it just a beautiful dream?

Delightfully enough, it can be done; and easily too. To do it, we begin by marking off logarithms on the slide rule.

For this all we need are the logarithms of the numbers from 1 to 10. Once we have these, we can obtain the logarithms of numbers less than 1 or more than 10 by choosing the proper characteristic, which is easy enough (see page 57).

The logarithm of 1 is 0.0. As we progress through numbers greater than 1, the logarithm increases through higher and higher values — 0.1, 0.2, 0.3, etc. — until finally the value of 1.0 is obtained, which is, of course, the logarithm of 10.

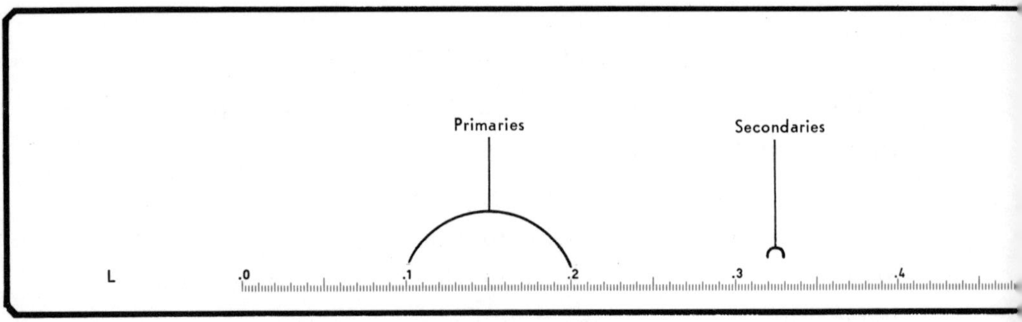

Suppose, then, we represent these logarithms in the form of a line on the body of the slide rule. The line begins at 0.0 (the logarithm of 1) and ends at 1.0 (the logarithm of 10). This line is divided into ten equal divisions by marks we can call *primaries.* The primaries represent logarithms of 0.1, 0.2, 0.3, and so on.

In order to simplify matters and keep from blurring the eye with unnecessary details, it is customary to omit the zero to the left of the decimal point. The primaries on this scale of logarithms are therefore marked off as .0, .1, .2, .3, .4, .5, .6, .7, .8, .9, 1.0, from right to left. In some slide rules the decimal point is also omitted.

This line actually deals with mantissas, rather than with logarithms, just as a logarithm table does. If we want to consider the line as dealing with mantissas to two places, the primaries mark off the mantissas 00, 10, 20, 30, etc. If we wish to consider the mantissas to three places, the primaries are 000, 100, 200, 300, etc.; if to four places, they are 0000, 1000, 2000, 3000, etc.

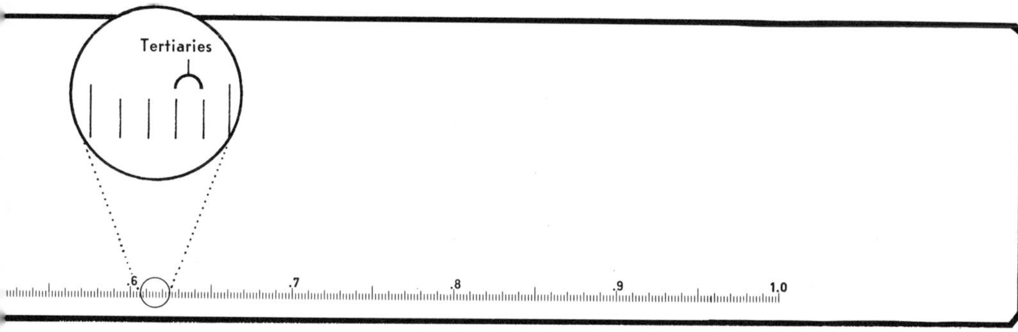

It will be most convenient to consider the mantissas to three places, and we will therefore consider the primaries as representing 000, 100, 200, 300, 400, 500, 600, 700, 800, 900, and 1000.

Between the primaries are ten equal divisions marked off by *secondaries,* the middle secondary being longer than the rest for convenience in reading mantissa values. Between each neighboring pair of secondaries are five still smaller divisions marked off by *tertiaries.*

The result, with its numbered primaries, and its unnumbered secondaries and tertiaries, is shown in Figure 24, in which the scale is marked "L" (for logarithm). We can, therefore, refer to it as the *L-scale.*

Notice that I have drawn the L-scale on the lower part of the body of the slide rule in Figure 24. There is no hard and fast rule about this. Whether the L-scale is drawn on the upper body or elsewhere depends on the particular slide rule.

Slide rules are manufactured in many different designs

which differ among themselves in details. The slide rule I am using as a model for the illustrations in this book has the L-scale on the lower body. Another slide rule in my possession has it on the upper body. Still others have it on the slide.

If your slide rule possesses scales in locations other than those shown in the illustrations, do not feel disturbed. The difference in location will not be important, and you will be able, very easily, to adapt my instructions to your slide rule.

If you have an inexpensive slide rule (and some quite decent slide rules can be obtained for not more than a couple of dollars) then some of the scales I will describe in this book may be missing altogether from your instrument. You will be able to follow the use of the missing scales from the diagrams in the book, however, and with the scales you do possess you will still be able to perform the essentials of multiplication and division.

If you measure the length of the L-scale on almost any slide rule, you will find that it is nearly ten inches long. It is this which allows people to refer to a 10-inch slide rule even though the overall length of the usual slide rule (which naturally extends some distance past both ends of the L-scale) is about 12½ inches long.

The fact that the L-scale has a total length of 10 inches means that the primaries (numbered from 1 to 10) are about 1 inch apart. This makes it tempting to look upon the slide rule as an "ordinary ruler." This, however, is quite wrong.

DO NOT make use of the L-scale to measure lengths. The slide rule is a precision instrument and should be used with the greatest care and only for the purpose for which it is designed. (Would you hitch a race horse to a milk wagon?) Secondly, the slide rule is not adapted to the purpose and has a blunt end against which it is hard to measure lines. Thirdly, the L-scale is not quite 10 inches long. It is 25 centimeters long and that actually comes to 9.84 inches. The primaries are therefore only 0.984 inches apart.

To summarize, the use of a slide rule as an ordinary ruler is improper, inconvenient, and incorrect. Need I say more?

It is not very difficult to read the L-scale. The reading can be made quite easily to three places so that the primary readings, as I said before, are 000, 100, 200, 300 and so on.

Concentrate for the moment on the primary markings 3 and 4 which represent 300 and 400. The secondaries lying between mark off tens: 310, 320, 330, 340, 350, and so on. The long secondary in the middle is 350. This can be told at a glance, and the shorter secondaries can be counted off easily enough.

Between the secondaries are the tertiaries which mark off by twos. Thus, between 330 and 340 are four tertiaries which represent 332, 334, 336, and 338. To find 337 we place an imaginary line midway between 336 and 338; and we can find in similar fashion any odd three-number mantissa.

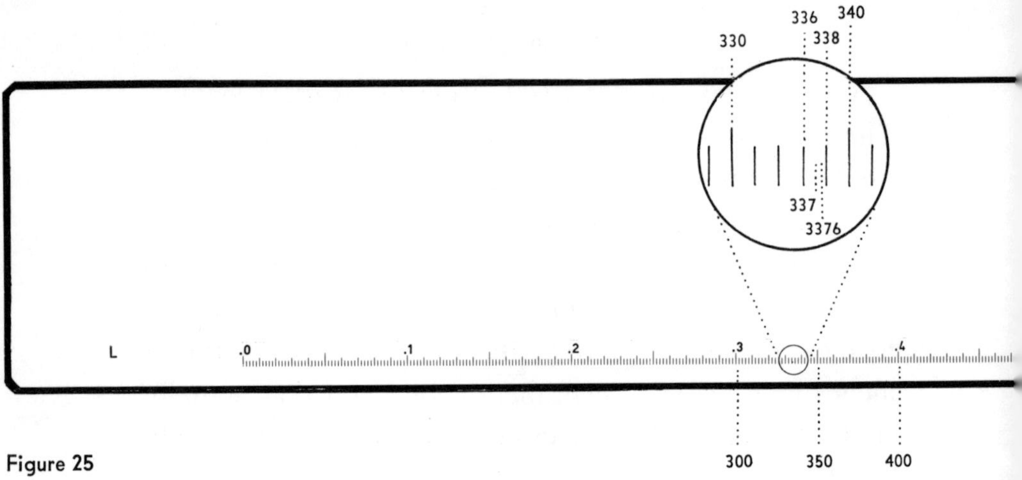

Figure 25

With practice one can even estimate four place mantissas. Suppose, having found 336 and 338, we consider those two tertiaries to represent 3360 and 3380. Exactly between lies the imaginary line marking 3370. Half way between 3370 and 3380 would be 3375 and a shade to the right of that would be 3376 (Figure 25).

Such an estimation of a fourth place is never accurate, however, and it is quite impossible to locate a fifth place on the L-scale. It is for this reason that the slide rule cannot entirely replace a logarithm table. A decent logarithm table gives logarithms to five places at least and thus offers an accuracy greater than that possible on the slide rule. The slide rule makes up for its lesser accuracy, however, as we shall see, by its far greater convenience.

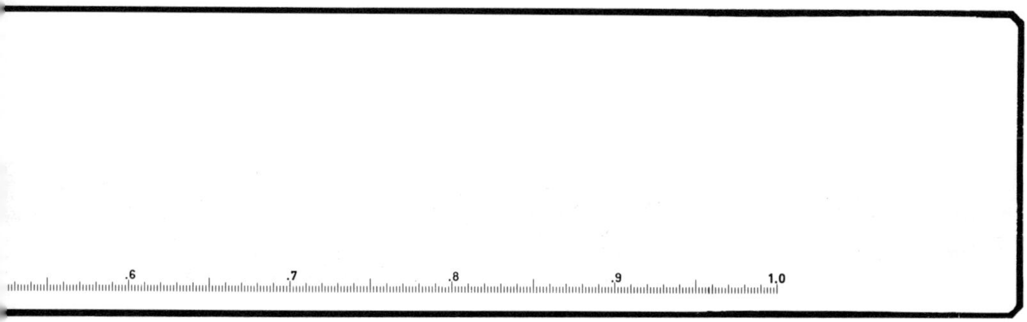

The D-scale

To make a logarithm table out of the slide rule, we need more than just the mantissas of the L-scale. We need antilogarithms, too, and we need those antilogarithms placed in such a way that we can in a moment match up each logarithm with its corresponding antilogarithm (or vice versa).

In order to do this we need a second scale above or below the L-scale. Each particular point on this second scale will represent the antilogarithm corresponding to the logarithm at the point directly above or below on the L-scale.

To match a point on one scale with the corresponding point on another scale directly above or below is not easy

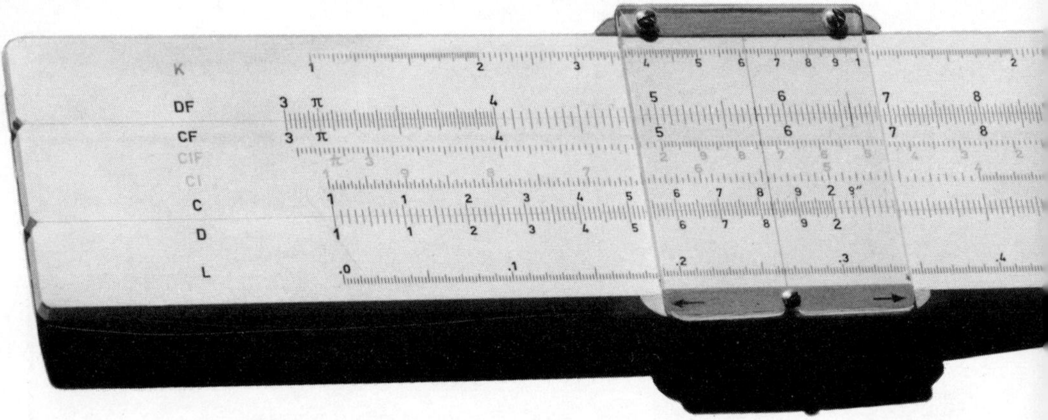

Figure 26

to do by eye, however. Slide rule manufacturers have therefore provided a mechanical device to help out. This is an *indicator assembly* which consists (in most slide rules) of two glass windows, one on either side of the slide rule, fitted into a holder at the top and bottom so that they are held firmly in place (Figure 26).

The indicator assembly can slide back and forth easily along the slide rule when it is pushed. If it is left alone, it is held in place by the friction of a spring, so that it stays where it is put. The indicator assembly cannot be moved off the slide rule, however, unless it is taken apart, for the metal bar that holds the two parts of the body firmly together acts to block the further movement of the indicator assembly.

Down the precise middle of each glass window is a fine vertical *hairline*. As the indicator assembly is moved right and left, the hairline itself can fall over any part of the L-scale from 0 on the far left to 10 on the far right.

The hairline can act as a convenient guide for the eye.

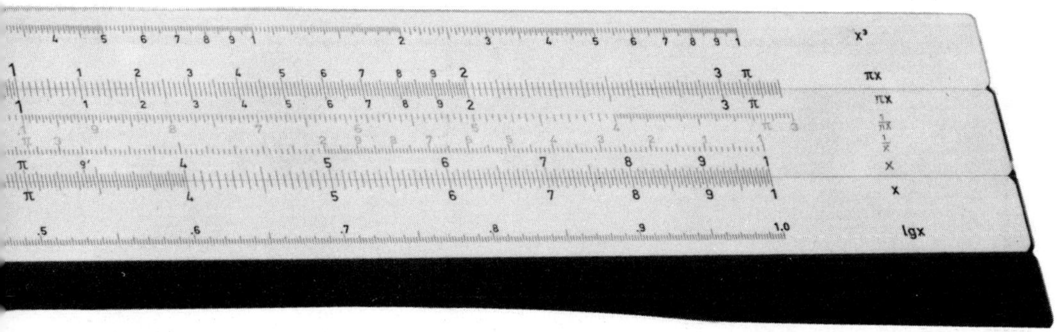

If you want to locate 337 on the L-scale (a reading we can refer to as L-337), begin by adjusting the hairline so that it falls between the tertiaries L-336 and L-338. It is easier to put a real line between the two marks, than to try to imagine one. By adjusting the hairline one can even try to estimate the position of L-3376 and mark its position until such time as the hairline must be moved again. The hairline, in short, finds a point on the scale and marks it.

It does more, too, for once the hairline is placed at L-337, it also marks the corresponding points on various scales above or below the L-scale. In particular, we can find points on the scale of antilogarithms (which I am about to discuss) that correspond to points on the L-scale.

The indicator assembly will play a part in virtually every drawing in the remainder of the book. In order to simplify these drawings and in order not to obscure any markings, I will make no attempt to picture the entire

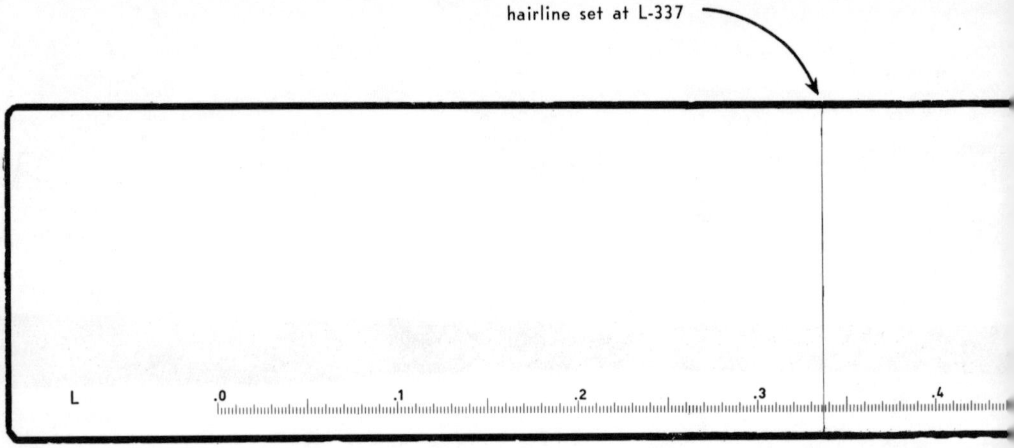

Figure 27

indicator assembly. I will represent only the hairline as a line crossing the slide rule vertically (Figure 27).

Where, now, shall we place our antilogarithm scale? There are four key scale positions on the slide rule: on the top and bottom of the slide, on the upper part of the body adjacent to the slide and on the lower part of the body adjacent to the slide. (These are the positions where I placed the four scales of my addition rule. See Figure 22 on page 38.)

Figure 28

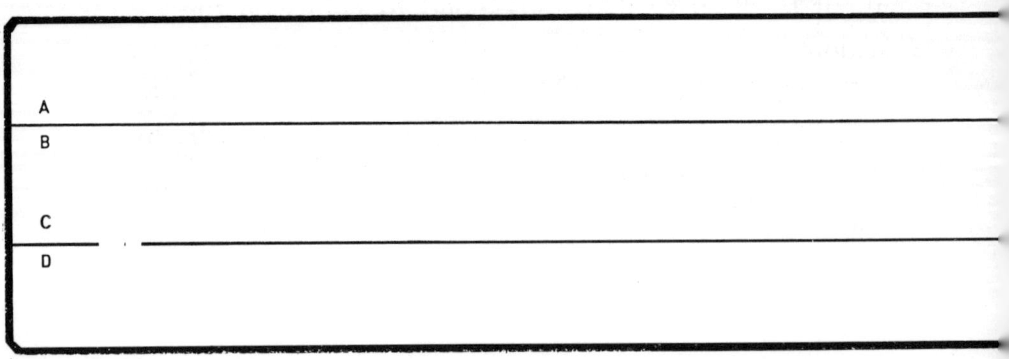

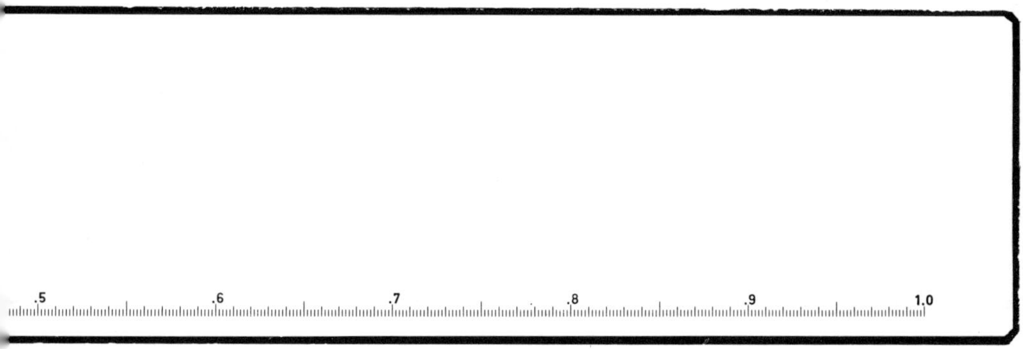

On early slide rules, these positions were used for the basic scales and they were marked, from the top down, A, B, C, and D (Figure 28). These letters are still applied to certain scales in these positions, although all other scales are lettered by means of initials of one sort or another, as the L-scale is lettered L for "logarithm."

It is customary to place one scale of antilogarithms in position D, and it therefore becomes the *D-scale*. By this arrangement, the L-scale and the D-scale are both on the

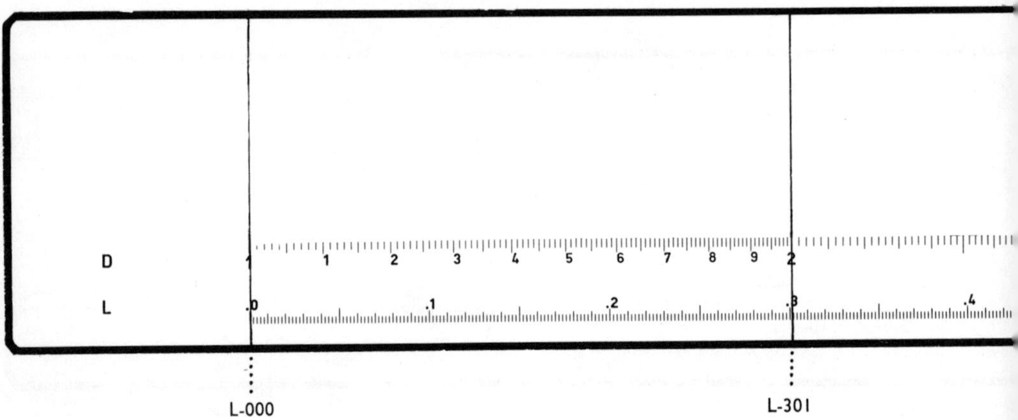

Figure 29

lower body. Each maintains a fixed and permanent position with respect to the other, since neither is disturbed by the to-and-fro motion of the slide.*

We can begin matching the L-scale and D-scale at the extreme left. The leftmost reading on the L-scale is 000, and this mantissa has for its antilogarithm 1, or any number obtained from 1 by moving the decimal point — 10, 100, 0.1, and so on. Naturally, it is simplest to consider the antilogarithm as 1.

If the hairline, therefore, is placed exactly on L-000, it should intersect the left end of the D-scale and give a reading of D-1.

(If, on your own slide rule, the hairline does not exactly intersect L-000 and D-1 simultaneously, then the slide rule is out of adjustment. It may be that the top and bottom halves of the body are not lined up properly, or

* Where the L-scale is on the slide, as in some rule designs, the scale of antilogarithms can be placed in the C-position. The L-scale and the C-scale will then have a fixed interrelationship.

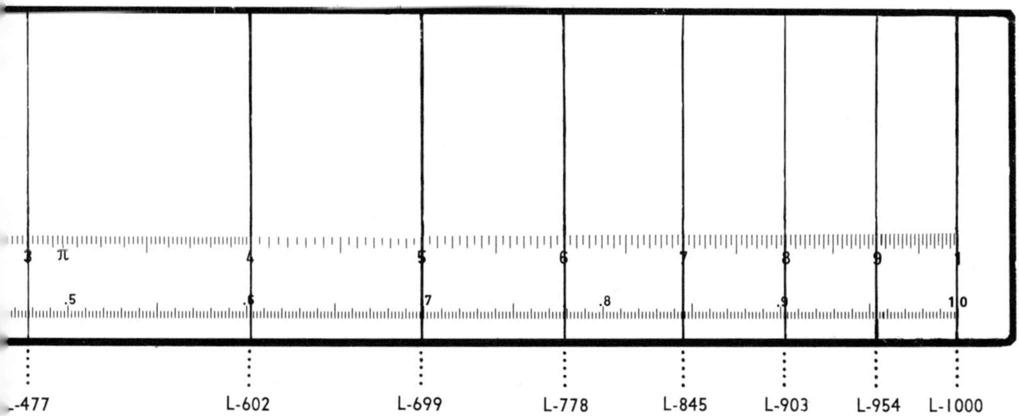

that the indicator assembly glass is not seated properly in its metal holder. It is best not to try the adjustment by yourself — a matter of loosening screws, moving the upper part of the body or the indicator glass, and tightening the screws again — but to allow someone familiar with the slide rule to do so.)

Let's continue. If we refer back to the table on page 54, we will see that *log* 2 = 0.301. (We need only use three decimal points here, for that is all we can read accurately on the L-scale.) That means *antilog* 0.301 = 2, so that immediately over L-301, we place D-2. Continuing to refer to the table on page 54, we must place D-3 directly over L-477, D-4 directly over L-602, and so on.

The reading L-1000 represents a logarithm of 1.000 and the mantissa here is once again 000. The antilogarithm is, once more, 1, so that the right end of the D-scale, like the left end, is marked 1. You see the result in Figure 29.

Compression at the Right

I'm sure that the first thing you notice about the D-scale is that the ten primaries marked off by the numbers are not equal in length. The distance between 1 and 2 is longer than the distance between 2 and 3, which is in turn longer than the distance between 3 and 4. The shortest distance of all, that between 9 and 10, is less than $\frac{1}{6}$ as long as the distance between 1 and 2. There is a steady compression of the distances at the right in other words.

To see why that is, let's consider the set of expressions: 2, 2×2, $2 \times 2 \times 2$, $2 \times 2 \times 2 \times 2$, and so on. If we consider the products of these expressions, we get the set of numbers: 2, 4, 8, 16, 32, etc., a set we have dealt with before.

Next let's consider the logarithms of the set of expressions we have just mentioned. The logarithm of 2 is 0.301. If we want the logarithm of 2×2, we must take the sum $0.301 + 0.301$. Continuing in this way, we find that the logarithms of the set of expressions 2, 2×2, $2 \times 2 \times 2$, $2 \times 2 \times 2 \times 2$, etc., is 0.301, $0.301 + 0.301$, $0.301 + 0.301 + 0.301$, $0.301 + 0.301 + 0.301 + 0.301$, etc. Taking the sums, we get the set of logarithms: 0.301, 0.602, 0.903, 1.204, and so on.

Now we can prepare a table matching the products of each expression in the set with the logarithm of that product, and we can begin with 1, which has the logarithm 0.000:

antilogarithm	*logarithm*
1	0.000
2	0.301
4	0.602
8	0.903
16	1.204

and so on for as long as we care to continue.

Notice that the antilogarithms increase by doubling. Each antilogarithm must be multiplied by 2 to get the next one in the set. A set of numbers in which each is obtained from the former through multiplication by a particular number (in this case 2) is a *geometric progression.*

The logarithms, on the other hand, increase by addition. To each logarithm must be added 0.301 to get the next one in the set. A set of numbers in which each is obtained from the former by the addition of a particular number (in this case 0.301) is an *arithmetic progression.*

It should be quite clear that the numbers in a geometric progression increase in value much more rapidly than do those in an arithmetic progression. You can see it's so in this case.

As we pass along the two sets of numbers in the table above, you can see that a fixed change in the logarithm brings about a greater and greater change in the antilogarithm. To begin with an increase of 0.301 brings the antilogarithm from 1 to 2; another such change of 0.301 brings the antilogarithm from 2 to 4; then from 4 to 8;

then from 8 to 16, and so on.

A broader and broader stretch of antilogarithms must fit into each fixed change of logarithm, which is why the numbers on the D-scale are closer and closer together as they get larger and larger. It is why there is compression on the right.

We can show this graphically, too, as in Figure 30. In this graph, we have a scale of logarithms from 0.0 to 1.2 on the bottom, and a scale of antilogarithms from 0 to 16 on the side. The dots mark the intersections of the logarithm value and its corresponding antilogarithm value, as given in the little table on page 77. A smooth curve is drawn through these dots and this represents the line of intersections of all logarithms with their corresponding antilogarithms.

As the logarithm increases at a steady rate (0.000, 0.301, 0.602, 0.903, 1.204), while the antilogarithm increases at a steadily increasing rate (1, 2, 4, 8, 16), the curve bends upward, becoming ever more nearly vertical.

If the horizontal lines, representing successive antilogarithms from 1 to 10, are reflected, so to speak, upward from the curve, we get a sequence of numbers from 1 to 10 at the top of the graph that are squeezed to the right exactly as they are on the D-scale. The antilogarithms, as positioned at the top of the graph, are exactly over their corresponding logarithms at the bottom of the graph. The top of the graph represents the D-scale, therefore, and the bottom of the graph the L-scale.

Since the compression at the right in the D-scale is

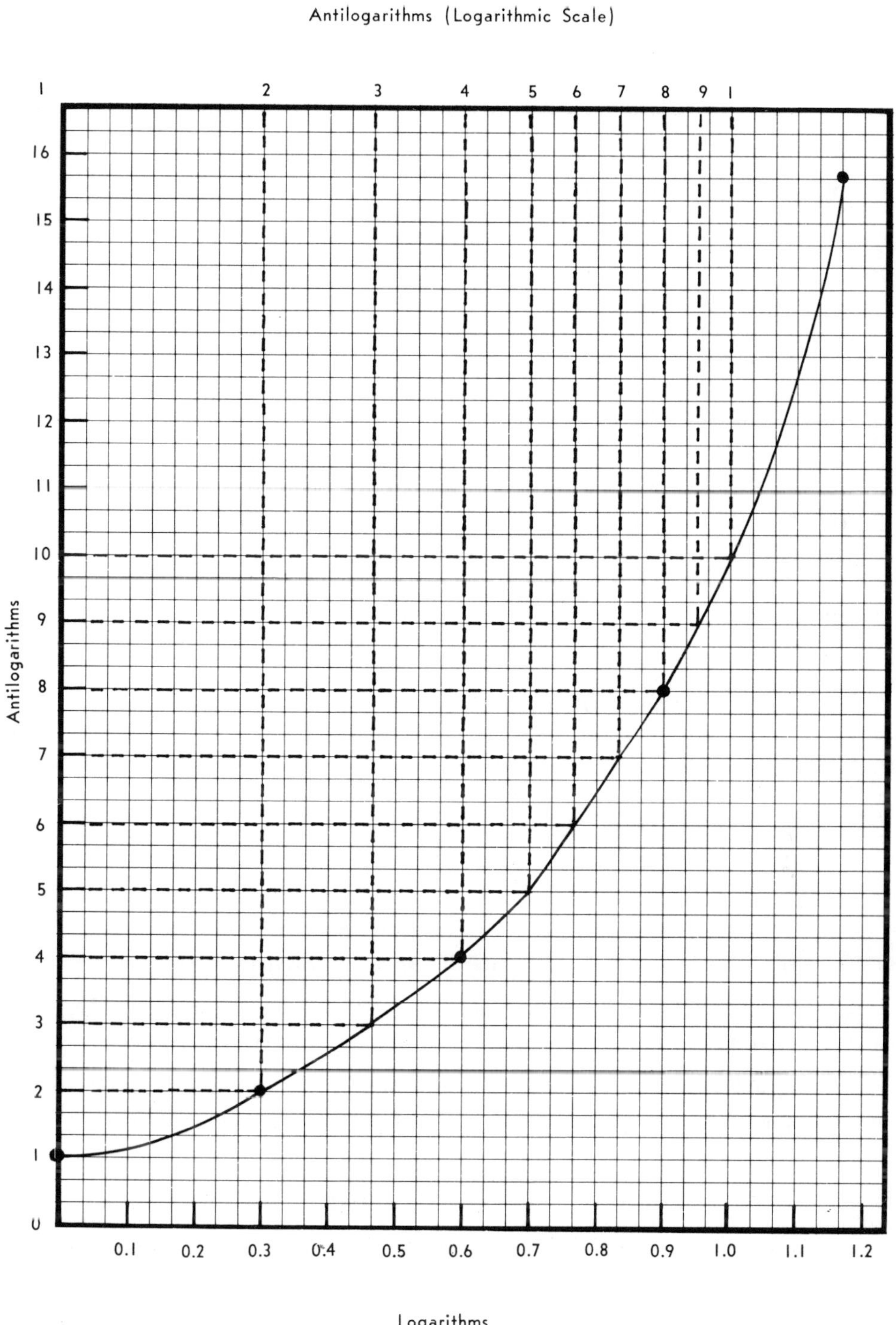
Antilogarithms (Logarithmic Scale)
1
2
3
4
5
6
7
8
9
1
16
15
14
13
12
11
10
9
8
7
6
5
4
3
2
1
0
Antilogarithms
0.1
0.2
0.3
0.4
0.5
0.6
0.7
0.8
0.9
1.0
1.1
1.2
Logarithms

Figure 30

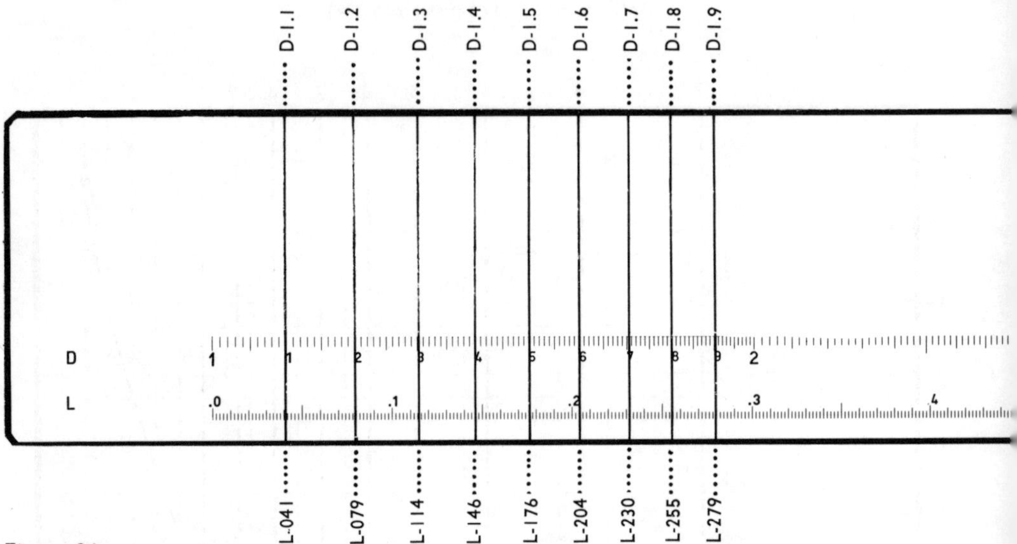

Figure 31

brought about by the adjustment of its numbers to a scale of logarithms increasing at a fixed rate, the D-scale is an example of a *logarithmic scale*.

Subdividing the D-scale

The spaces between the primaries on the D-scale can be divided into finer subdivisions. Consider the space between D-1 and D-2, for instance. D-1 is under L-000 and D-2 is under L-301 because $log\ 1 = 0$ and $log\ 2 = 0.301$. Using the same system we can place the values 1.1, 1.2, 1.3, and so on, onto the D-scale directly over appropriate values on the L-scale.

Thus, $log\ 1.1 = 0.041$; $log\ 1.2 = 0.079$; $log\ 1.3 = 0.114$, etc. Therefore, D-1.1 can be placed over L-041; D-1.2

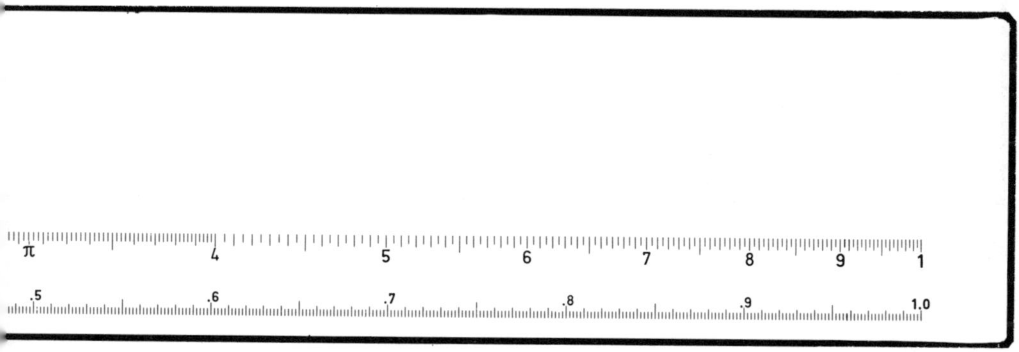

over L-079; D-1.3 over L-114, and so on (Figure 31). This gives us our secondary markings between D-1 and D-2. Similarly, tertiaries can be placed between D-1.1 and D-1.2, marking off 1.11, 1.12, 1.13, and so on. In this way, too, secondaries and tertiaries can be set up all along the length of the D-scale as finely and as accurately as the length of the slide rule and the delicacy of the instruments used will allow.*

The usual secondaries and tertiaries on the D-scale of an ordinary slide rule are shown in Figure 32. You will notice two things. First, the secondaries and tertiaries get more crowded as one moves to the right. This is to be expected in a logarithmic scale. The second is the con-

* The difficulty of making an accurate logarithmic scale contributes to the expense of a good slide rule.

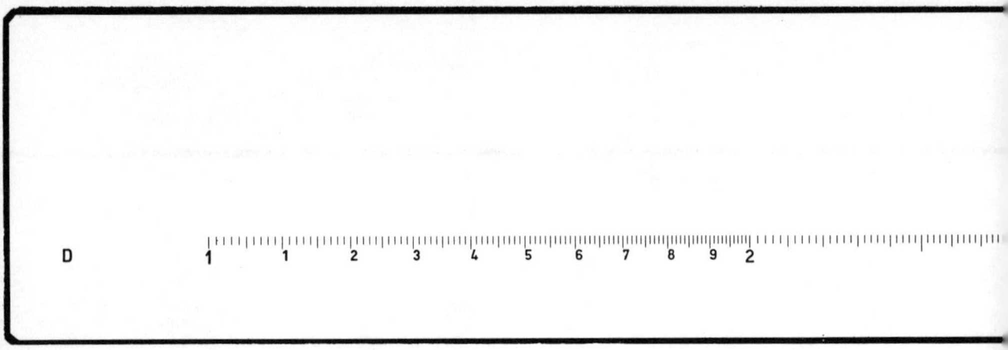

Figure 32

sequence of this, for the tertiaries represent different values in different portions of the scale.

This should not be surprising. In an ordinary scale such as the L-scale, the spacing between the primaries is equal throughout and secondaries and tertiaries can be distributed evenly all along the line. In a logarithmic scale, however, the space between the primaries steadily decreases as we move to the right so that room can be found for fewer and fewer subdivisions.

As it happens, secondaries representing 0.1 each are distributed all along the D-scale but get progressively closer, and the space between them available for tertiaries gets progressively skimpier. Consequently, there are tertiaries of three different kinds present on the D-scale.

Consider first, the space between D-1 and D-2, which takes up nearly a third of the total length of the D-scale. It can be divided up finely. The secondaries, representing tenths, are so relatively far apart that it is hard for the eye to take them in at a glance. For that reason (and

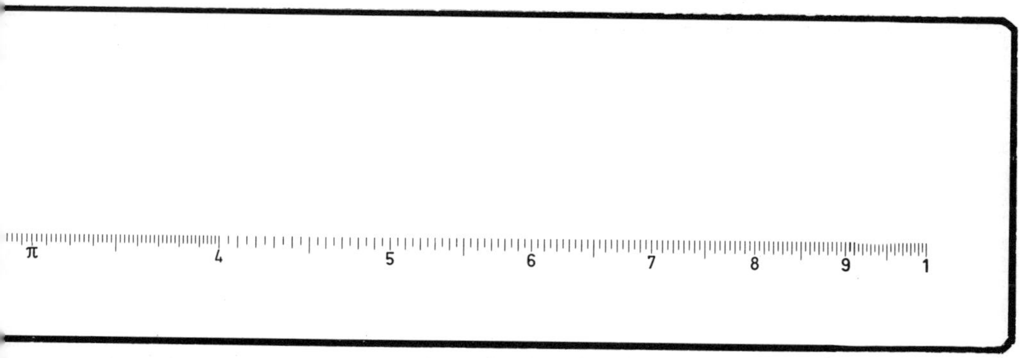

because there is enough room for it), the secondaries between D-1 and D-2 are marked with the numbers 1 through 9, numbers of smaller size than those marking the primaries (see Figure 32). The small number 1 marks the 1.1 value, the 2 marks the 1.2 value, and so on up to the 9, which marks the 1.9 value.

The spaces between these secondaries is divided into ten smaller divisions. Thus, there are tertiaries, each representing an increase in value of 0.1, running from 1.01 for the first tertiary after D-1 to 1.99 for the last tertiary just before D-2. The middle tertiary between neighboring secondaries, marking numbers such as 1.05, 1.15, 1.25, and so on, is longer than the others.

In addition, one can place the hairline midway between two tertiaries so that one can indicate 1.005 or 1.235 without trouble, or make rough estimates for 1.007 or 1.233.

Beyond D-2, the secondaries do not possess numbers actually marked on the scale. If numbers were present they would be crowded too closely for clarity. On the

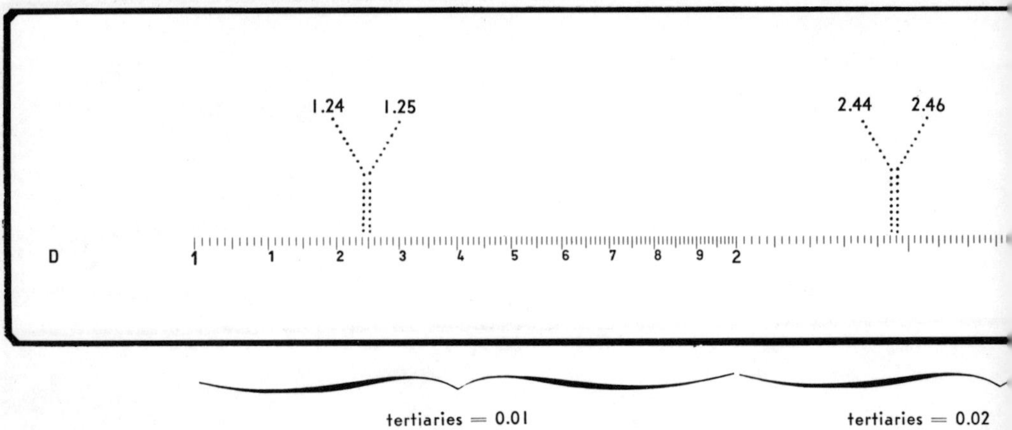

Figure 33

other hand, the middle secondary in each case, representing 2.5, 3.5, 4.5, and so on, is made longer than the rest, and using these as guides, the value of any other secondary is quickly seen.

Since the stretch from D-2 to D-4 is just equal to that from D-1 to D-2, there are twenty secondaries crowded into the former stretch, where only ten exist in the latter. For that reason, there is no room in the D-2 to D-4 stretch for the secondaries to be conveniently divided into ten smaller divisions. Five smaller divisions are set up instead and each tertiary in this stretch therefore represents a value of 0.02. Thus, between D-2.1 and D-2.2 are tertiaries representing 2.12, 2.14, 2.16, and 2.18. By centering the hairline between such tertiaries, one can get 2.13, 2.15, and so on.*

* Near the tertiary 3.14 is the special marking π. It is not present in all slide rule designs. Its use will be discussed on page 111.

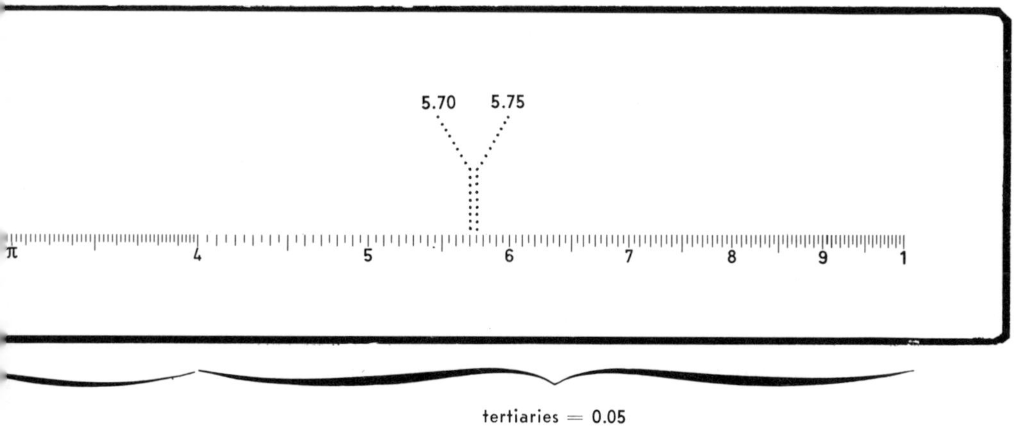

The stretch from D-4 to D-10 is divided into sixty secondaries that crowd closer and closer together. So close are they that each space between secondaries in this portion of the D-Scale is divided merely into two smaller divisions, with a single tertiary representing a value of 0.05. Thus, between D-4.3 and D-4.4 is a tertiary indicating 4.35, and between 9.6 and 9.7 is a tertiary indicating 9.65. By centering the indicator between a tertiary and a secondary, one can obtain numbers such as 4.325 or 9.675.

To summarize, we can divide the D-scale into three segments in which the value of the tertiaries differs (see Figure 33). From D-1 to D-2, each tertiary represents a value of 0.01; from D-2 to D-4, a value of 0.02; and from D-4 to D-10, a value of 0.05.

Naturally, when one is not accustomed to a logarithmic scale, this change in the nature of the tertiaries can be confusing. There will be some stumbling at first, inevitably, but continued use of the slide rule will eventually make it all seem natural and there will be no trouble in reading the D-scale, no more than in reading the markings of an ordinary ruler.

It might seem to you from all this that the left end of the D-scale is more delicate than the right end, since the left end has tertiaries that represent finer divisions. Let's look into that situation.

Suppose you are trying to read the hairline position on the left end of the scale, and find it just about halfway between the tertiaries 1.34 and 1.35. You conclude that the reading, therefore, is 1.345. However, it may seem to you that the position is just a shade to the right of the midpoint, so that the reading might be 1.346. You can't really tell exactly which of these it is so you have an uncertainty of 0.001.

On the right-hand side of the scale you might make a reading of 9.475, because the hairline is just halfway between the tertiary marking 9.45 and the secondary marking 9.50. However, again you cannot be sure that it is exactly in the center. The space between 9.45 and 9.50 (a difference of 0.05) is only about two-thirds as wide as that between 1.34 and 1.35 (a difference of only 0.01) and its exact midpoint is harder to judge. If the hairline seems a bit to the right of the midpoint, the reading can easily be 9.485 rather than 9.475, an uncertainty of 0.01.

Since the uncertainty at the left is 0.001 and the uncertainty at the right is 0.01, you might think that the accuracy at the left end is ten times as great as that at the right end. That, however, is not so.

What counts is not the size of the uncertainty in itself, but its size compared to the size of the number being measured. What we ought to consider is the uncertainty percentage. Thus, if you consider an uncertainty of 0.001 in a number such as 1.345, the uncertainty is about 0.075 percent of the number. An uncertainty of 0.01 in a number like 9.47 represents an uncertainty that is about 0.105 percent of the number.

These percentages are not very different. We can say, with reasonable confidence, that the accuracy of a slide rule is the same all along its length and that it is everywhere easily reliable to about a tenth of one percent, or one part in a thousand. (A very careful user could squeeze out an even better accuracy.)

Using the Wooden Log Table

Let's consider the L-scale and the D-scale taken together. What can be done with them? Obviously, the two scales have been constructed in such a way as to give us a three-place logarithm table, with the logarithms in the L-scale and the antilogarithms in the D-scale.

Suppose you want the logarithm of 3.5. You place the hairline over D-3.5 and find that it also marks L-544

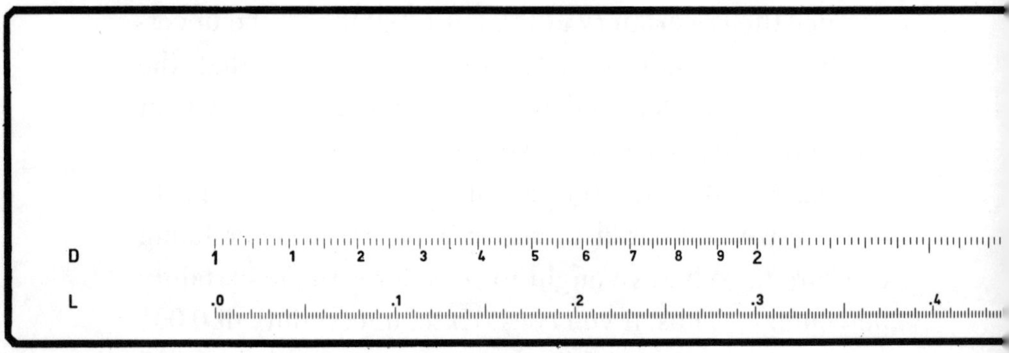

Figure 34

(Figure 34). You can conclude then that *log* 3.5 = 0.544. If you were to look up the logarithm of 3.5 in a five-place logarithm table, you would find that *log* 3.5 = 0.54407 so that your slide rule answer is certainly satisfactorily close, particularly since all you need to do to get it on the slide rule is to shift the hairline. It is a much simpler task to do that than to leaf through an ordinary logarithm table and run your finger down columns of numbers.

Of course, the L-scale gives you only the mantissa, but that is all you need. If you wanted the logarithm of 35, or 350, or of 0.35, you would still place the hairline on D-3.5 to get L-544. You then merely adjust the characteristic according to the principles described on page 54. We then see that *log* 35 = 1.544, *log* 350 = 2.544, *log* 0.35 = 0.544 − 1, and so on.

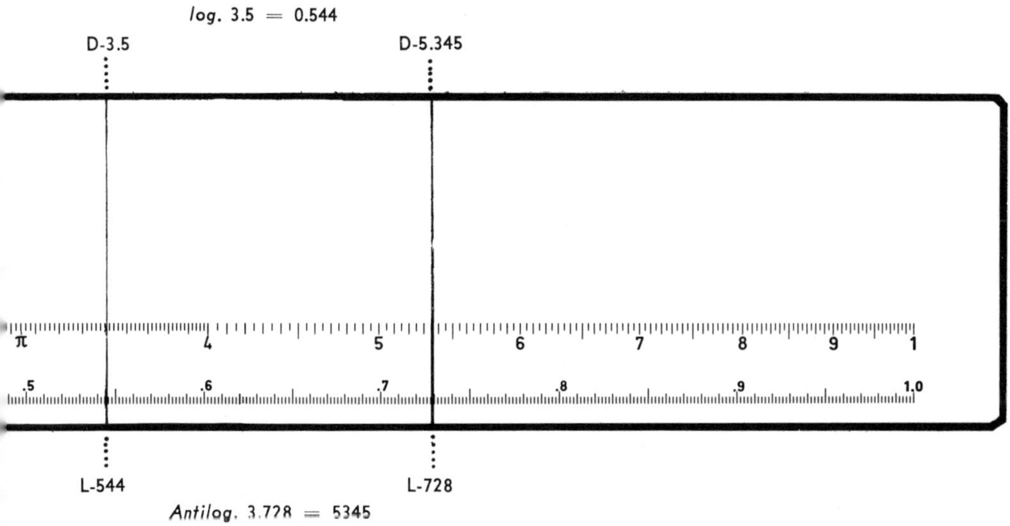

It works the other way round, too. Suppose you are given the logarithm 3.728 and want the antilogarithm. To use the L-scale, which gives only the mantissa, you drop the characteristic temporarily and search for 728 only. You place the hairline at L-728 and find that the D-reading is just a shade to the left of 5.35 (see Figure 34). You estimate the reading to be 5.345. The characteristic tells you where the decimal ought to go. Since the characteristic is 3, the number of digits to the left of the decimal point must be $3 + 1$, or 4. Consequently, *antilog* $3.728 = 5345$. A five-place logarithm table will tell you that *antilog* $3.728 = 5345.4$.

As you see, then, the D-reading is not fixed. What we read as 3.5 can be 350; and what we read as 5.345 can be 5345 or 0.0005345, for that matter. A reading on the D-scale gives us a fixed set of digits, but it says nothing about the position of the decimal point. That belongs to us.

For this reason, I will no longer try to place a decimal point in the D-reading. At first, it may seem very convenient to consider the secondary marks as representing "tenths" and to read them as 1.1, or 2.4 or 9.6, but I will consider such readings as D-11, D-24, or D-86 respectively. In the same way tertiary marks will not represent 1.46 or 2.46 or 9.55. They will be D-146, D-246, and D-955 respectively.

You may feel lost without the decimal point just at first, but you will get used to that too. And the loss is only temporary. As you will see, once a slide rule computation is completed, in will go that decimal point.

5

Multiplication

The C-scale

WE HAVE now converted the slide rule into a logarithm table and, having done so, we find that we don't really need to labor in order to use it. The slide rule makes its use automatic. Whenever the hairline is placed on any D-reading, it simultaneously marks off the logarithm of that number as the L-reading. We don't have to look for the logarithm; it is there.

This helps us understand a crucial difference between the evenly spaced scale on an ordinary ruler (such as the scales we used to work our addition rule in Chapter 2) and the logarithmic D-scale.

In an ordinary scale, the numbers mark off lengths that correspond in value to the numbers themselves. The number 2 is two inches from the left end; the number 5 is five inches from the left end and so on. By adding and subtracting these lengths, therefore, we add and subtract numbers (see page 5).

The numbers on the D-scale, however, mark off lengths that correspond not to the numbers themselves, but to the logarithms of the numbers. On the L-scale, the

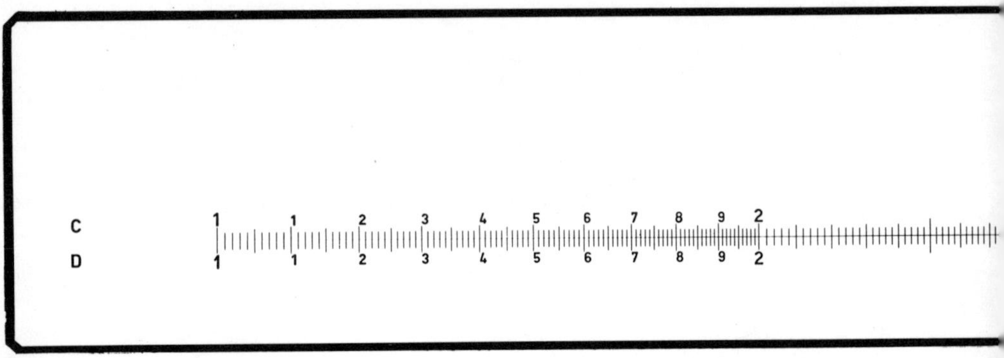

Figure 35

primaries are usually placed about an inch apart (see page 66). That means that since D-1 is placed over L-0 and D-2 is placed over L-301, D-2 is about 3 inches from the left end of the scale. Similarly, since D-3 is under L-477, D-3 is about 4¾ inches from the left end of the scale. These distances, I repeat, are not equivalent to the numbers themselves but to the logarithms of the numbers.

Consider, then! When we used an addition rule with its scales representing lengths equivalent to the numbers upon it, we could manipulate those lengths so as to add and subtract numbers. But if we use something like the D-scale in which the numbers represent lengths equivalent to the logarithms of those numbers, we can manipulate those lengths so as to add and subtract logarithms. In adding and subtracting logarithms, we are, of course, multiplying and dividing the antilogarithms — that is, the numbers on the D-scale.

To perform such manipulations on the addition rule,

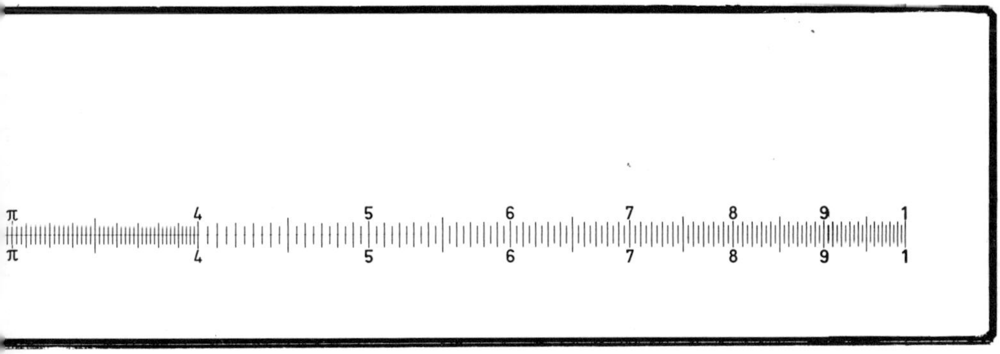

we needed two scales of identical construction that could be moved against each other. We need the same now — two scales of identical construction that can be moved against each other.

Therefore, on the bottom of the slide, in the C-position (see page 73), another scale, exactly like the D-scale, is constructed. This new scale, the *C-scale,* can be placed exactly over the D-scale, with every marking on the former over the corresponding marking on the latter and we can then say the slide rule is in a neutral position (Figure 35). The C-scale can, however, be shifted out of neutral and moved along the D-scale in such a way as to add and subtract logarithms.

We can begin with a very simple: 2×3.

To do this, we follow a similar procedure to that used in adding numbers on the addition rule (see page 16). We begin by moving C-1 over D-2.

And this at once brings up an important point. There are two C-1's, one at the left end of the scale and one at

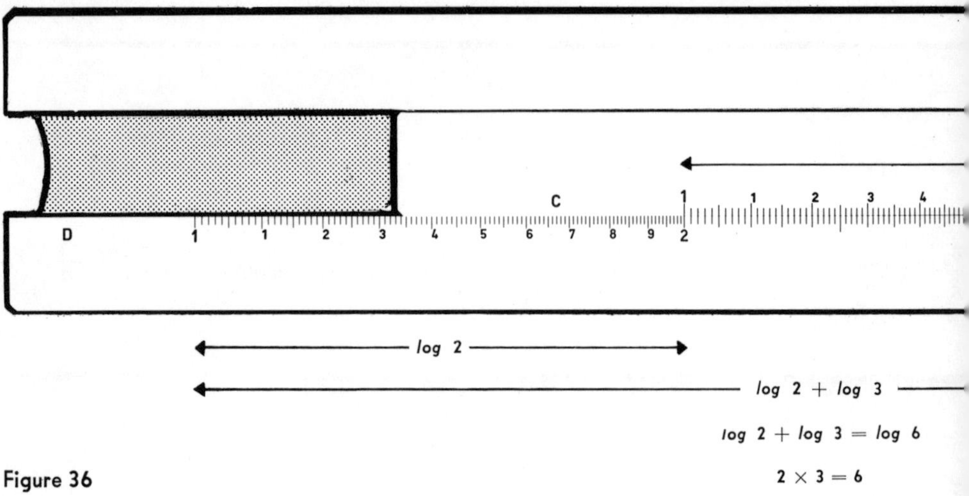

Figure 36

the right end. These are used so often in slide rule manipulations that they are often referred to as the *left index* and the *right index,* respectively. In this book, however, in order to avoid confusion, I will give no readings any special name. I will refer to the 1 at the left end of the C-scale (the more frequently used) as C-1. The 1 at the right end of the C-scale, I will refer to as C-1-(right). In the same way there is both a D-1 and a D-1-(right).

In placing C-1 over any D-reading, it is not necessary to use the hairline as a guide. Standing as it does at the very beginning of the scale, C-1 is clearly marked out and the eye can follow it without trouble. Markings along the interior of the scale are easy to confuse with their neighbors, however. Readings between markings

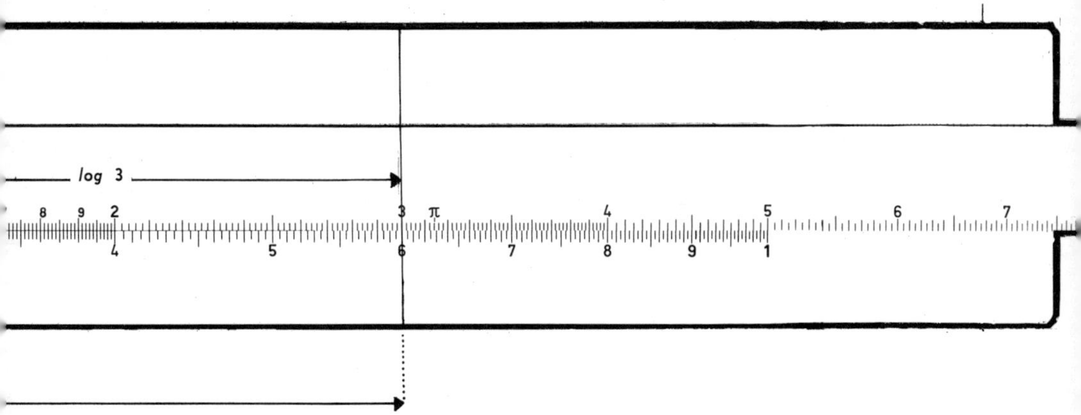

are even easier to confuse. It is routine, then, to place the hairline over any reading one wishes to make that does not involve the end-marking of a scale.

With all this in mind, we can return to the multiplication 2×3. Carrying it through by the system we used in the addition rule, we place C-1 over D-2, then shift the hairline over to C-3 (Figure 36).

Since D-2 marks off a distance of *log* 2 from the end of the D-scale, and C-3 marks off a distance of *log* 3 from the left end of the C-scale, we are adding *log* 2 and *log* 3 in this manipulation. Since *log* 2 + *log* 3 = *log* 6 (and you can check this in a log table), we would expect to find D-6 under C-3, for D-6 would be a distance of *log* 6 from the end of the D-scale.

We do not, however, bother reading the slide rule in

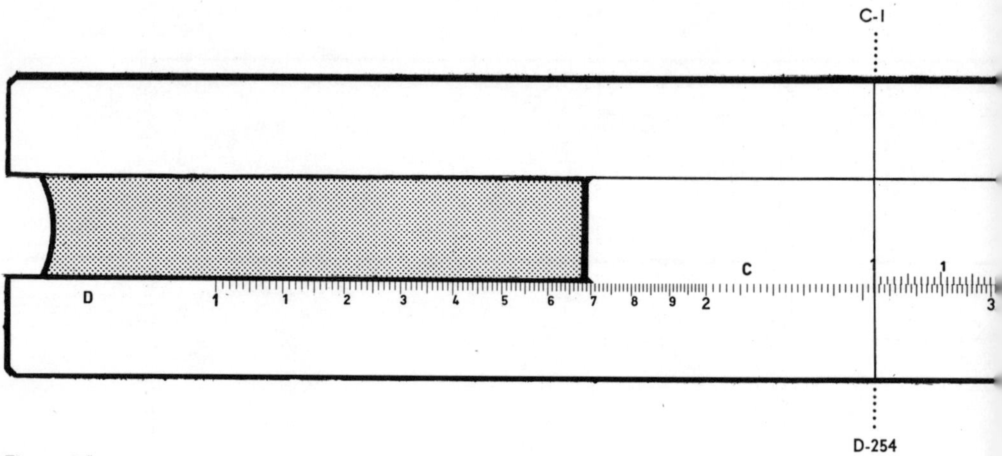

Figure 37

the position shown in Figure 36 as an addition of logarithms. We do not say: *log* 2 + *log* 3 = *log* 6. The D-scale gives the antilogarithms and we read those antilogarithms directly, converting the addition of logarithms into a multiplication of antilogarithms. We say, $2 \times 3 = 6$.

The same thing would happen if we wanted to multiply 3×2. We would place C-1 over D-3, then move the hairline to C-2, and find D-6 immediately under C-2. This is not surprising for *log* 2 + *log* 3 = *log* 3 + *log* 2, and $2 \times 3 = 3 \times 2$.

This is simple so far. We scarcely need a slide rule to tell us that $2 \times 3 = 6$. However, what if it is 2.54×3.76 that we are interested in. The multiplication is now

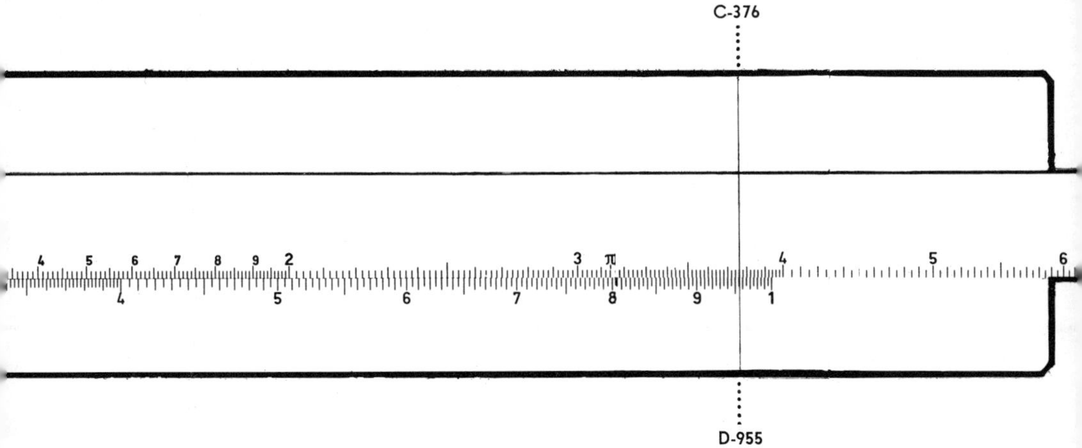

2.54 × 3.76 = 9.55

much more difficult on paper, but not a bit more difficult on the slide rule.

Move C-1 over D-254 (taking a little care to make sure you have the right tertiary), and then move the hairline to C-376. You find the hairline to be over D-955 (see Figure 37) and conclude that 2.54 × 3.76 = 9.55. Work it out in full and you will find that the correct product is 9.5504, but surely 9.55 is close enough — and think of the saving in time.

But what if it were 2.54 × 4.76 you wanted to solve? You begin, again, by moving C-1 over D-254, but this time you find you are stuck, for the C-476 which you must reach with your hairline is off the D-scale and can give you no D-reading.

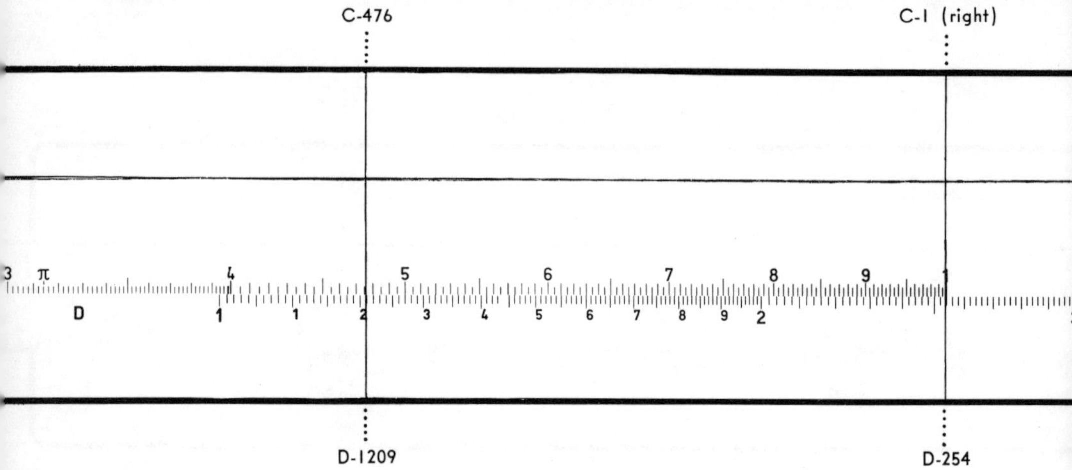

2.54 × 4.76 = 12.09

Ah, but we went through that in the addition rule (see page 19) and the same device will serve us here. We make use of the C-1(right) in place of the C-1 and put that over D-254. Now we can find C-476 over the D-scale. The nearest tertiary is C-475 so we place the hairline just a bit to the right of that, about one fifth of the way over to C-480. (Naturally, this means a kind of "judging by eye" but this can be done pretty well, especially with practice.) The hairline, on C-476, also falls just to the left of D-121 (which can also be read as D-1210). We might judge the mark to be D-1209 (Figure 38), and conclude that $2.54 \times 4.76 = 12.09$. If we work the multiplication in full, the product turns out to be 12.0904.

The Decimal Point

This brings up the question of the decimal point. In the example just completed (2.54×4.76) the slide rule

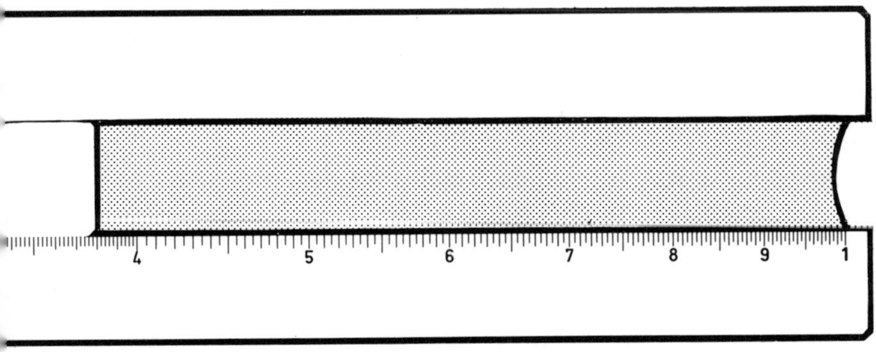

Figure 38

gave us the digit-combination 1209 as the answer. We might have decided that this meant $2.54 \times 4.76 = 1.209$, or that $2.54 \times 4.76 = 120.9$, but we didn't. We said that $2.54 \times 4.76 = 12.09$. Let's see why.

Consider the following multiplications:

$$1.5 \times 3.23 = 4.845$$
$$15 \times 32.3 = 484.5$$
$$0.015 \times 0.323 = 0.004845$$
$$150{,}000 \times 0.00323 = 4845$$
$$150 \times 323 = 48{,}450$$

This shows us once again that the decimal point is a side issue. If we go by the slide rule we deal only with digit-combinations and all the examples above boil down to $15 \times 323 = 4845$. The decimal point — that side issue — is for us to handle.

There is, however, no need to feel aggrieved, for, though placing the decimal point may be a tedious necessity, it is not difficult. Occasionally, a bit of carelessness will result in a misplaced decimal point, but this can happen even in pencil-and-paper calculations. The correct response to carelessness is a sober determination to be careful, that's all!

The best way to place the decimal point is to consider the problem and substitute, for the numbers involved, similar numbers that are particularly easy to handle. Such similar numbers will give you an answer that is wrong, of course, but one that is close enough to the right answer to have the decimal point in the same place. You will have an answer that is of the same *order of magnitude.*

Each shift of a decimal point by one place produces a change of one order of magnitude. For instance, 2.54 is one order of magnitude smaller than 25.4 and two orders of magnitude smaller than 254. Again, 2.54 is one order of magnitude larger than 0.254 and two orders of magnitude larger than 0.0254.

If two numbers differ by a factor of less than 5, they can be lumped together as of the same order of magnitude. Thus, 17 and 68 are of the same order of magnitude since 68 is only four times as great as 17. On the other hand 17 is not of the same order of magnitude as 2.14 since 17 is eight times as great as 2.14. However, 17 is of the same order of magnitude as 21.4 which is, in turn, one order of magnitude greater than 2.14. Therefore, we can say that 17 is one order of magnitude greater than 2.14.

Now let's look at one of the problems in the list given

on page 99; say, 15×32.3. We can convert 15 to 20 and 32.3 to 30. The numbers are changed but not the order of magnitude. It is easy to multiply 20 by 30 in our head. The answer is 600. That is not the answer we are looking for, but it is the same order of magnitude as the answer. If the slide rule tells us that the digit-combination of the product is 4845, then to make that the same order of magnitude as 600, we must write it 484.5. We know then that $15 \times 32.3 = 484.5$.

A similar treatment will give us the correct answer for any other example in the list.

And it will also give us the correct answer to the problem with which I started this section, 2.54×4.76. If we make the first number a little higher, changing it to 3, and the second a little lower, changing it to 4, we retain the order of magnitude. Since we know that $3 \times 4 = 12$, we know that the slide rule answer of 1209 must be written 12.09 to keep the product in the correct order of magnitude. Therefore, $2.54 \times 4.76 = 12.09$ and nothing else.

Practice is all you need to learn to handle orders of magnitude almost automatically, and once you've worked that out in advance, you can manipulate the slide rule, get your digit-combination, and place your decimal point without any hesitation.

Let's take another example, and a slightly more difficult one.

Suppose you wanted to carry through the following: $2.72 \times 7.23 \times 1.15 \times 0.86$. This represents three multiplications to be carried out one after the other.

You can begin by placing C-1 over D-272 but then you will find that C-723 is off the D-scale. You therefore "switch indices," bringing C-1(right) over D-272, and carrying the hairline to C-723. (To find C-723, you must place the hairline between the secondary representing 720 and the tertiary representing 725 — and place it a little closer to the 725 than to the 720.)

Having done this, you will find that the hairline is also marking out a reading just under D-1970. You don't, however, have to try to estimate what the answer is, or even look at it. The hairline marks the product of 2.72 × 7.23 on the D-scale and keeps marking it as long as you leave the hairline in place. This product must next be multiplied by 1.15.

You therefore bring the C-1 to the hairline (making sure you don't move the hairline position in the process) and this places C-1 over the D-reading that is the product of 2.72 × 7.23. To multiply that product by 1.15, you move the hairline (which you no longer need in its first position now that the C-1 marks it) to C-115. That marks a new D-reading which represents the product of 2.72 × 7.23 × 1.15.

This new product must be multiplied by 0.86. You therefore bring C-1 to the hairline marking that product and find that C-86 is off the D-scale. You bring C-1-(right) to the hairline instead and then move the hairline to C-86.

Now the hairline marks out a D-reading just about midway between D-194 and D-195. Call it D-1945. That is the digit-combination that represents the product of 2.72 × 7.23 × 1.15 × 0.86 (Figure 39).

But where is the decimal point? If we look at the multiplication problem again, we can change 2.72 to 3, 7.23 to 7, 1.15 to 1, and 0.86 to 1. The problem becomes 3 × 7 × 1 × 1 which you can see at a glance equals 21. To give 1945 the same order of magnitude as 21, we must write 1945 as 19.45. Therefore 2.72 × 7.23 × 1.15 × 0.86 = 19.45.

(I have, by the way, just worked out the triple multiplication by pencil and paper as quickly as I could and it took me two full minutes to get the answer 19.4492784. I worked it out slowly and carefully by slide rule and it took me just 15 seconds — one-eighth the time — to get 19.45, which is almost exactly the correct answer.)

Folded Scales

Sometimes, as you saw in the case discussed just above, you must use C-1 and sometimes C-1(right). With practice, you get the "feel" for which one to use.

Suppose, for instance, you consider 2.14 × 37.6. You can adjust these figures to their easy approximations of 2 and 40 and see, at a glance, that 2 × 40 = 80. In that case, you will recognize that it is safe to bring C-1 to D-214 for you expect C-376 to be somewhere near D-8,

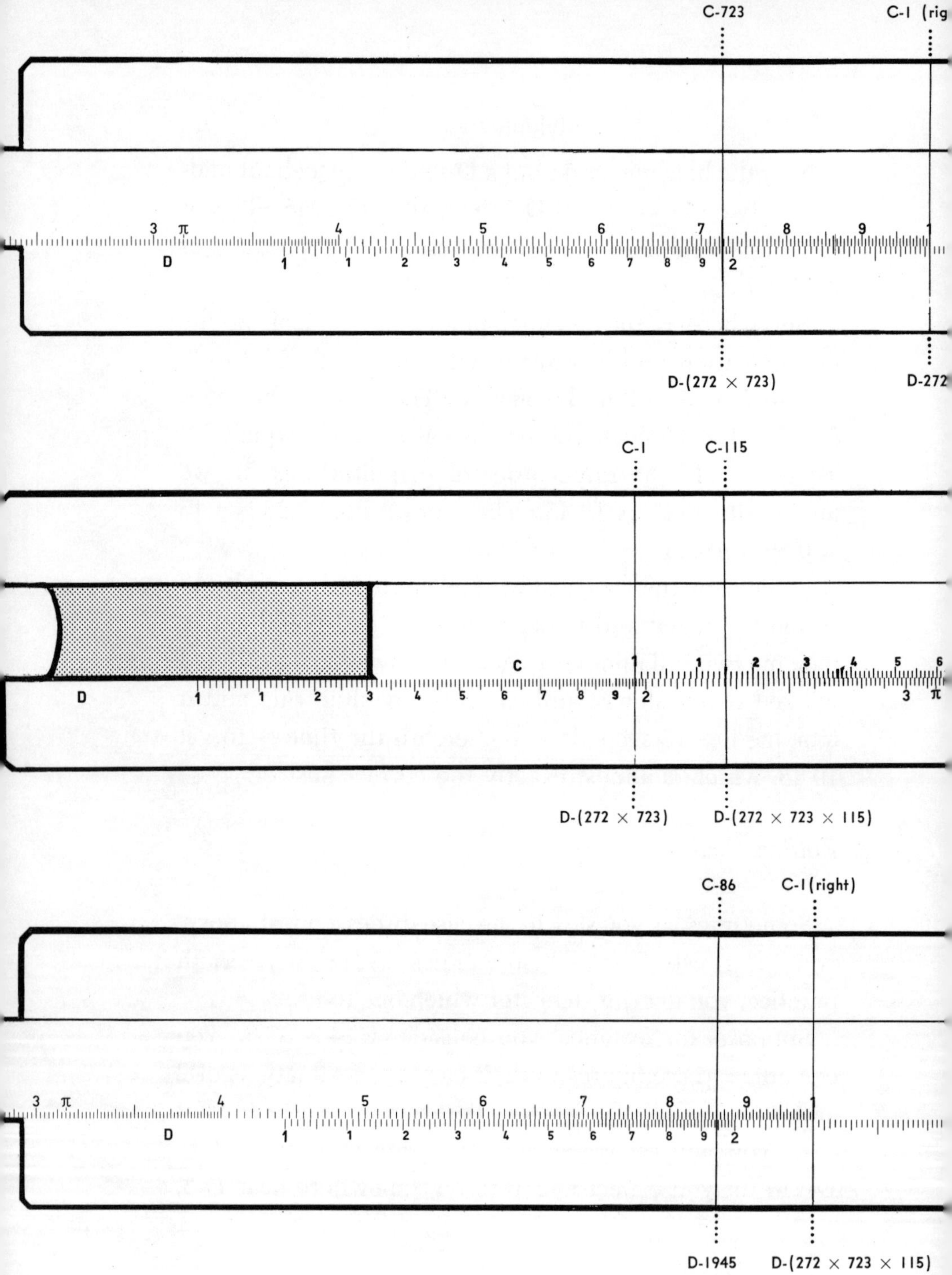

Figure 39

$2.72 \times 7.23 \times 1.15 \times 0.86 = 19.45$

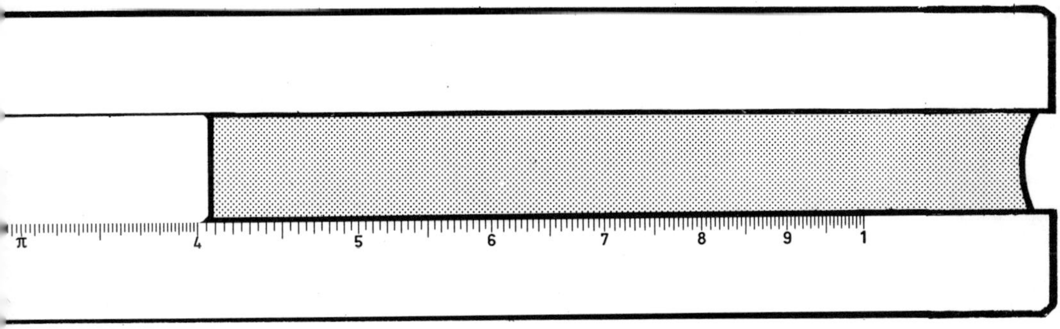
π 4 5 6 7 8 9 1

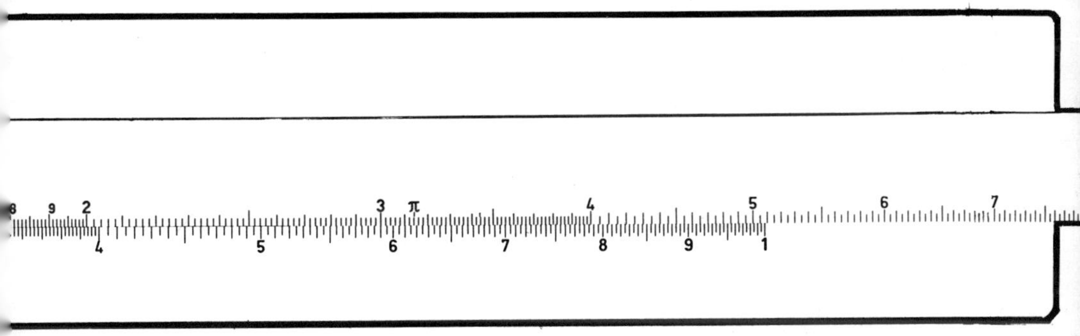
8 9 2 3 π 4 5 6 7
4 5 6 7 8 9 1

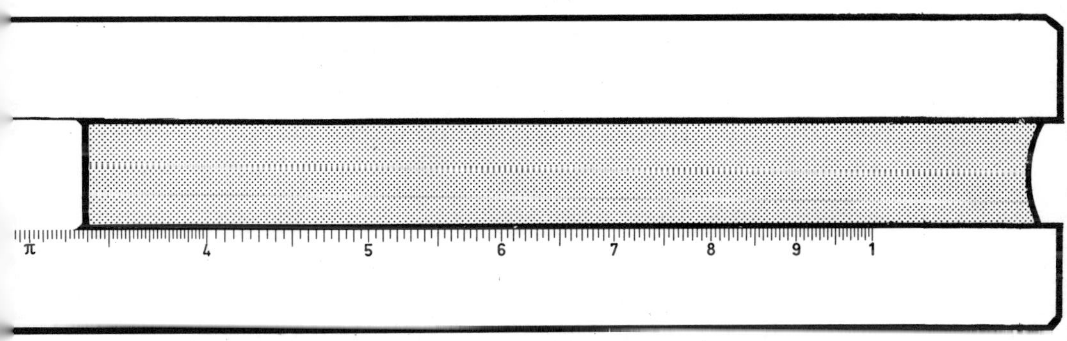
π 4 5 6 7 8 9 1

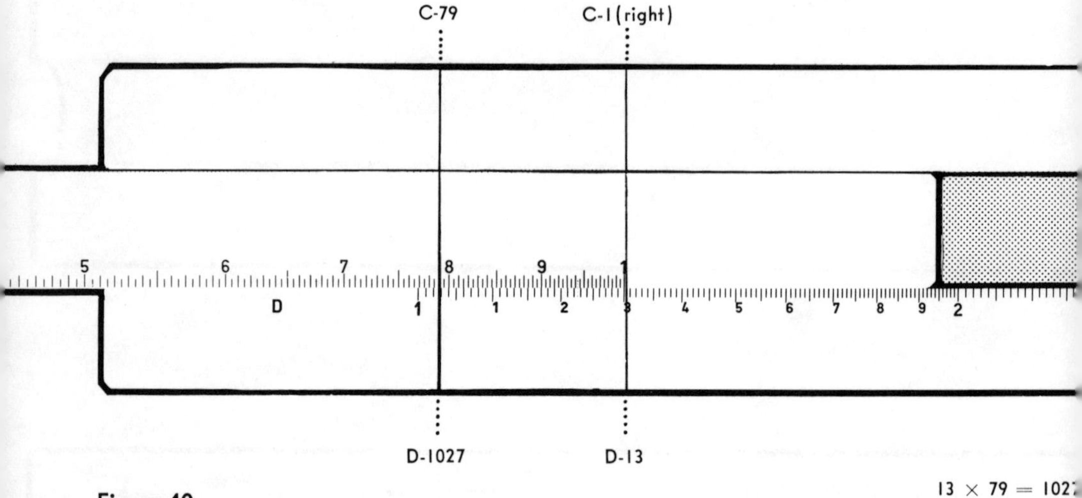

Figure 40

which would keep it safely on the D-scale. Sure enough, if you perform the manipulation, you will find C-376 over D-805 so that 2.14 × 37.6 = 80.5.

If, on the other hand, you were dealing with 4.14 × 3.76, which is close to 4 × 4, which in turn is equal to 16, you would expect the answer to be beyond D-1 (right), which would represent a digit-combination of only 10. You can consider D-1 itself to be 10 also, of course, but in that case 16 would be near the left end of the D-scale and must be sought there. Therefore you bring C-1-(right) to D-414 and then find C-376 over D-1556 so that 4.14 × 3.76 = 15.56.

However, it is always possible to be fooled. Suppose you try 13 × 79 and decide to approximate it at 10 × 80 = 800. Therefore you move C-1 over to D-13 expect-

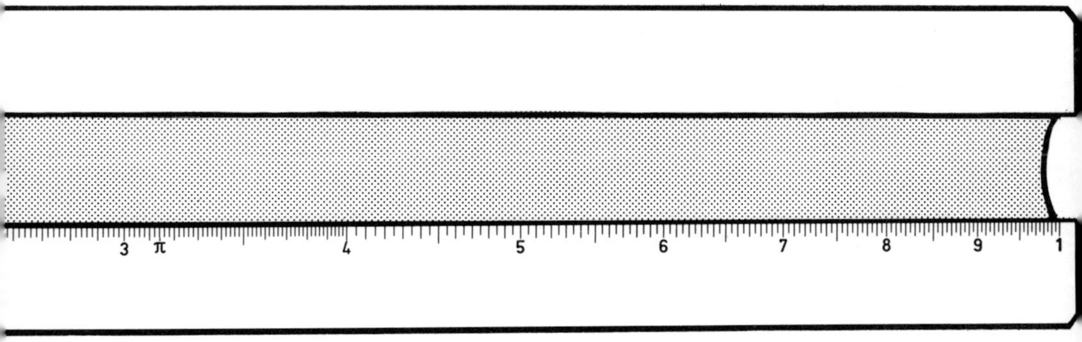

ing to find C-79 somewhere near D-8 and safely on scale. You find instead, to your horror, that C-79 is just beyond the end of the D-scale. You must therefore switch indices and bring C-1(right) all the way over to D-13 and then find C-79 over D-1027 (Figure 40) and decide that $13 \times 79 = 1027$.

Whenever this happens, there is bound to be a certain amount of muttering under the breath.

It would be nice if this could be avoided. Suppose, for instance, that the D-scale were continued past D-1-(right) so that there were a D-2(right), a D-3(right) and so on. Then, if a C-reading moved off the ordinary D-scale, it could be picked up on the D-scale extension. The disadvantage of this is that the slide rule must be lengthened, making it both more expensive and more

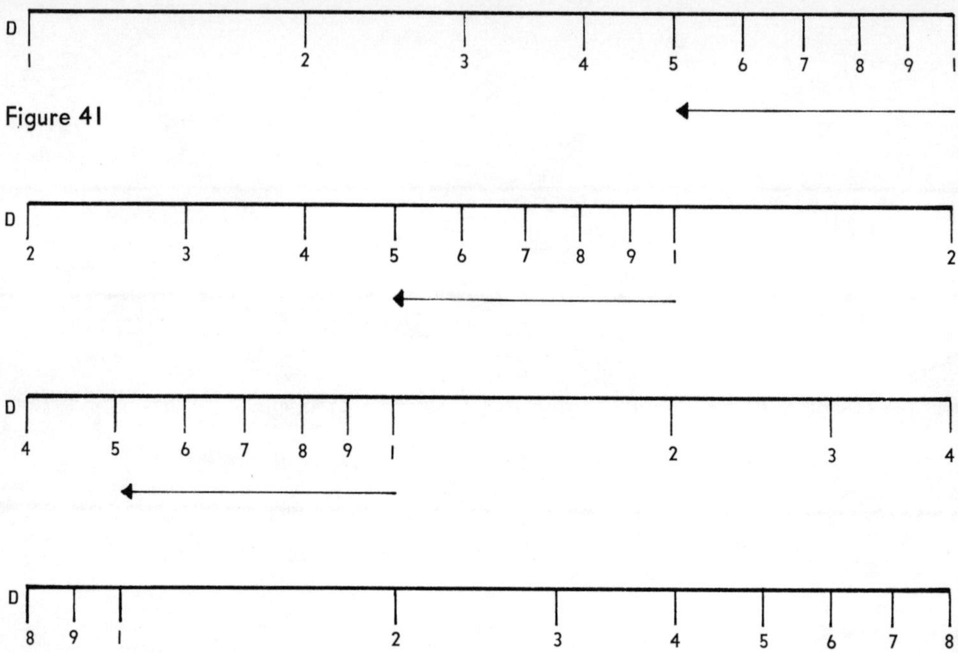

Figure 41

unwieldy. (In some slide rule designs, however, particular scales are extended about half an inch or so past either end to take care of borderline cases.)

Another way out is to move the D-scale bodily to the left so that D-1(right) is moved leftward, leaving room beyond it for an extension without making the slide rule any longer in that direction.

If you keep the rest of the D-scale intact, however, it then begins to stick out at the left end like a tusk. In order not to extend the slide rule in either direction, we must arrange to have the left end of the D-scale disappear as it moves leftward. The additional room at the right for an extension must then be made up for by a disappearance of the scale at the left (Figure 41).

As you move D-1(right) leftward — provided you keep the overall length of the scale fixed — the extension you gain on the right is exactly balanced by the loss on the left. It follows then, as you can see in Figure 41, that the reading on the left end of the scale is always the same as on the right end of the scale. If the scale begins with D-2, it ends with D-2(right); if it begins with D-4 or D-8, it ends with D-4(right) or D-8(right).

As you move D-1(right) leftward, you experience a gain because of the extension on the right, and you also experience a loss because of the disappearance on the left. To increase the gain as much as possible you want to push D-1(right) continually leftward. To minimize the loss you want to push D-1(right) continually rightward. The two impulses balance, as you might expect, precisely midway and the ideal situation is to have D-1 exactly in the middle of the scale.

If D-1 were exactly in the middle of the scale what would be the numbers on the left end and right end? (Whatever number would be at the left end would, of course, also be at the right end.)

To answer this question, let us remember that the entire stretch of the D-scale represents the stretch of the L-scale from 0.000 to 1.000 (see page 74). If D-1 is in the middle of the scale, the logarithmic stretch from that middle to the right end is, obviously, 0.500 or, in fractions, ½. Earlier in the book, I pointed out that $10^{1/2} = \sqrt{10}$, or 3.16. Therefore, antilog 0.500 = 3.16.

The numbers on the D-scale are placed at the distance

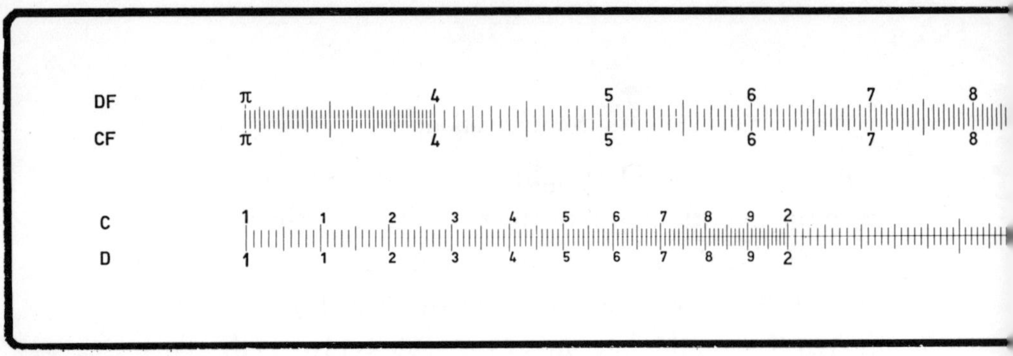

Figure 42

from D-1 that corresponds to the length of the logarithm of those numbers. Therefore the right end of a D-scale that has D-1 in the middle must be 3.16, since the distance of 3.16 from D-1 is then just half the length of the scale or 0.500 (in logarithms) and that is, indeed, the logarithm of 3.16. Consequently the left end of the D-scale with D-1 in the middle must be 3.16 also. Omitting decimal points, a D-scale with D-1 in the middle should run from D-316 to D-316(right).

Such a scale is called a *folded scale,* the notion being that an ordinary scale, folded in half, has 316 appearing at the fold, and the new scale starts and ends at that fold.

In my slide rule, the folded scales are present in the A- and B-positions (see page 73), but they are not referred to as A- and B-scales. Those names are reserved for other scales which I will discuss later in the book.

Instead, the scale in the A-position is the *DF-scale,* while the scale in the B-position is the *CF-scale* (Figure 42). The DF-scale is so named because, like the D-scale,

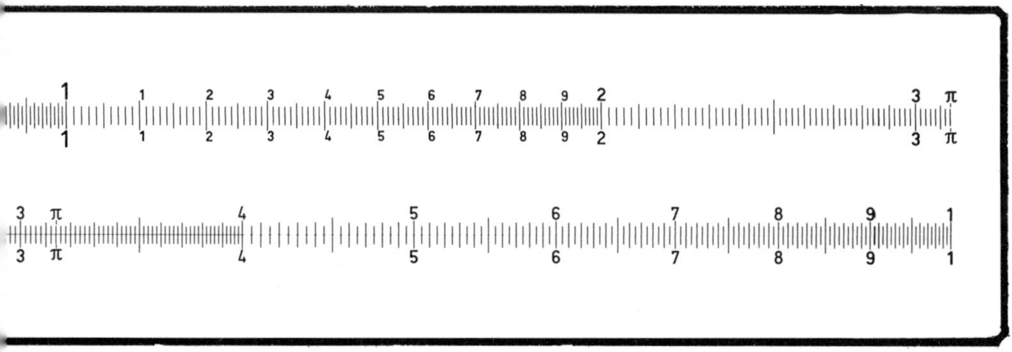

it is on the body, while the CF-scale, like the C-scale, is on the slide. The F, of course, stands for "folded."

Pi

If you look at Figure 42 carefully (or, better still, look at a slide rule) you will see that the DF-scale and the CF-scale do not actually start and end with 3.16 but with 3.14 and that the two ends are marked thus: π.

Behind that lies a story. In the case of every circle, however large or small, the length of the circumference is about $3\frac{1}{7}$ times that of the diameter. The actual ratio is a never-ending decimal, but, to five decimal points, it is 3.14159 which, as you see, is not far removed from 3.14. This ratio is customarily symbolized by the Greek letter π ("pi").

It so happens that π is not confined merely to questions involving the length of the circumference of a circle in terms of its diameter. It crops up in numerous mathematical equations and comes into play in almost every facet of science.

Consequently, scientists and engineers frequently find occasion to multiply π by some number. That is the reason for the special π marking on the C- and D-scales. It is also the reason why the DF- and CF-scales are designed to start with π. Although those scales should, ideally, start with 3.16, it is not much of a departure from the ideal to have them start just to the right of 3.14 instead. It is an alteration of not quite a single tertiary and the insignificant bit of imperfection introduces such a great deal of convenience that no one would object.

To see what this means, let's begin by noting that C-1 is directly under CF-π. This is a fixed position, for both the C-scale and the CF-scale are on the slide and neither can be moved relative to the other. This is also true for the D-scale and DF-scale which are both on the body so that one can't be moved relative to the other. Therefore, assuming both parts of the body to be in perfect adjustment, D-1 is permanently under DF-π.

Next, let's consider the hairline at some point on the C-scale — any point. Let us say the reading is C-*a*. At the same time, the hairline is indicating a reading on the CF-scale; say, CF-*b*.

We already know that the C-*a* reading is at a distance from the left end of the C-scale equivalent to the logarithm of *a*. Since the left end of the C-scale is C-1, which represents a logarithm of 0.000, the logarithm of the C-*a* reading is 0.000 + *log a,* or simply *log a.*

The CF-b which is simultaneously marked off by the hairline also represents a logarithm, which we can call *log b*. But CF-b (marked out by the hairline simultaneously with C-a) must be as far from the left end of the CF-scale as C-a is from the C-scale. Hence CF-b is a distance from CF-π that is equivalent to *log a*. But CF-π is equivalent to log π. Therefore, log $b = log\ \pi + log\ a$. If we convert those logarithms into antilogarithms (and remember to change the addition into a multiplication in so doing), we find that $b = \pi \times a$ (Figure 43).

If this is not instantly clear to you, it will become clear as soon as you put the notion into actual practice.

Suppose you want the value of 2π. You have only to move the hairline to C-2, and you will find the hairline will simultaneously give a reading of CF-6.28, which is equal to 2π.

In the same way you can find with a single setting of the hairline that $4.3\pi = 13.51$, since, when the hairline indicates C-43, it also indicates CF-1351. As other examples you can find that $75.2\ \pi = 236$, that $0.388\pi = 1.22$, and so on. The placing of the decimal point is no problem, for you have but to remember that multiplying by π is not much different from multiplying by 3.

Avoiding the Shift

But multiplication by π is a side issue. When I started discussing the folded scales, it was not with such a multi-

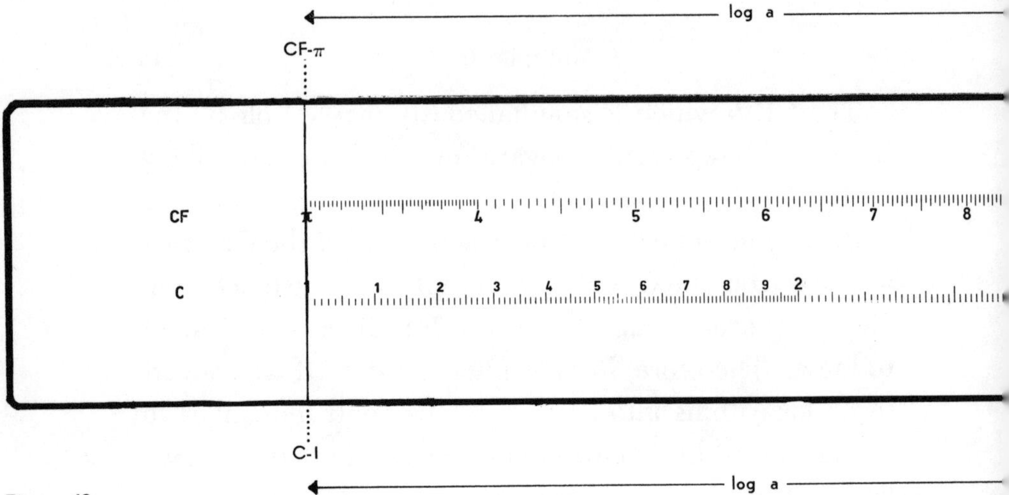

Figure 43

plication in mind. The folded scales were introduced in order to make it unnecessary to shift from C-1 to C-1-(right) in ordinary multiplications. (Or at least to reduce the number of times the shift must be carried through, if the necessity isn't eliminated altogether.)

In order to go back to this earlier problem, take another look at the CF/DF scales as compared with the C/D scales. When the slide rule is in its neutral position, C-1 is over D-1, and CF-1 is under DF-1, as shown in Figure 42.

If the slide now moves in such a way as to displace the C-scale aganst the D-scale, it also displaces the CF-scale against the DF-scale. And since the C-scale and the CF-scale are on the same piece of wood, the amount of displacement must be precisely the same for both.

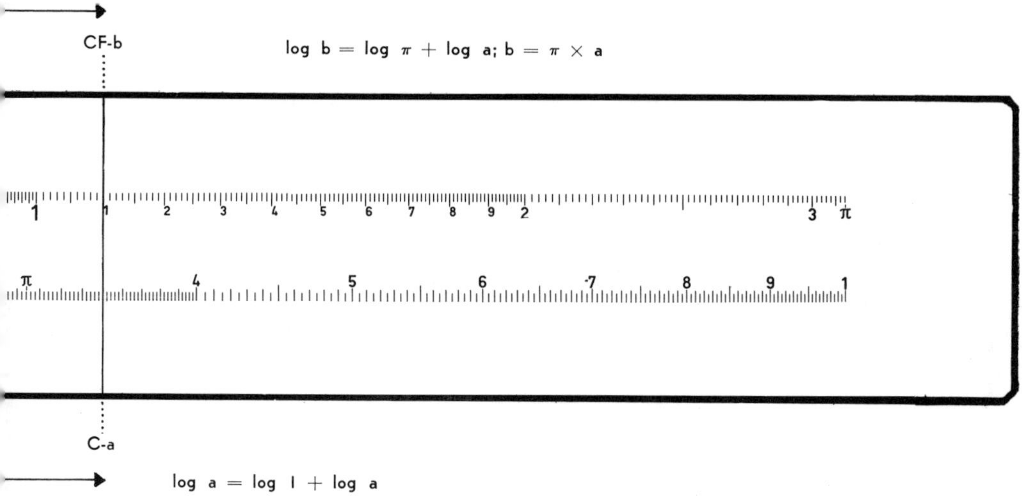

Suppose, for instance, that you move C-1 over D-226. You will find that CF-1 will move directly under DF-226 (Figure 44). Furthermore, if you look at other portions of the scales, you will see that there are similar situations all along the line. Thus, C-13 is directly over D-294 and CF-13 is directly under DF-294, and so on for any other readings you care to make.

Since the readings on the C- and D-scales are exactly duplicated by the readings on the CF- and DF-scales, it doesn't matter which pair of scales you use for multiplying. You can use the folded scales exclusively if you wish; or you can move freely from the ordinary scales to the folded scales and back again.

There is an advantage to moving back and forth between the two sets of scales. Suppose C-1 is placed over

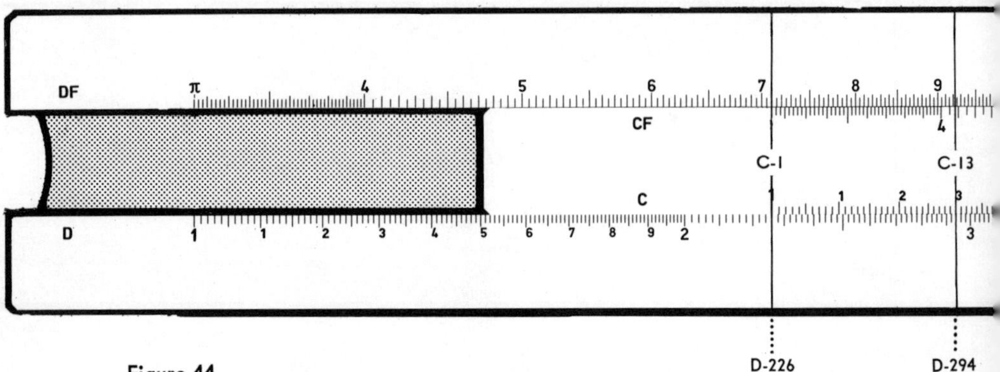

Figure 44

D-314. In that case, half the C-scale moves beyond the right end of the slide rule and the markings from C-314 into the higher numbers can't be used. Half the CF-scale also moves beyond the right end of the slide rule, but here it is the numbers smaller than CF-314 that can't be used (Figure 45).

The situation is, then, that numbers up to 314 are usable on the C-scale and numbers over 314 are usable on the CF-scale. On one scale or the other, all the numbers are usable even though half the slide is beyond the right

Figure 45

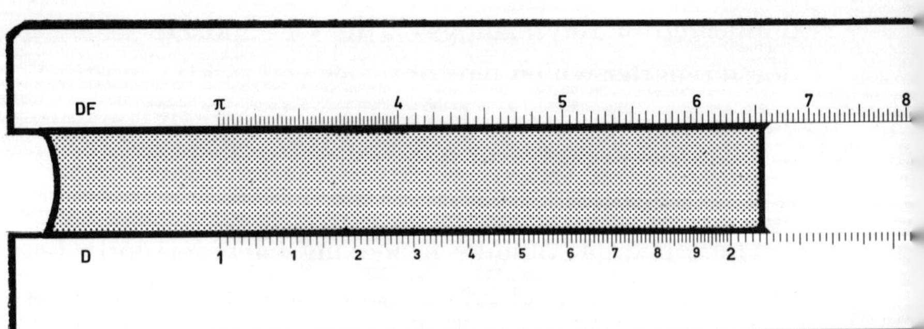

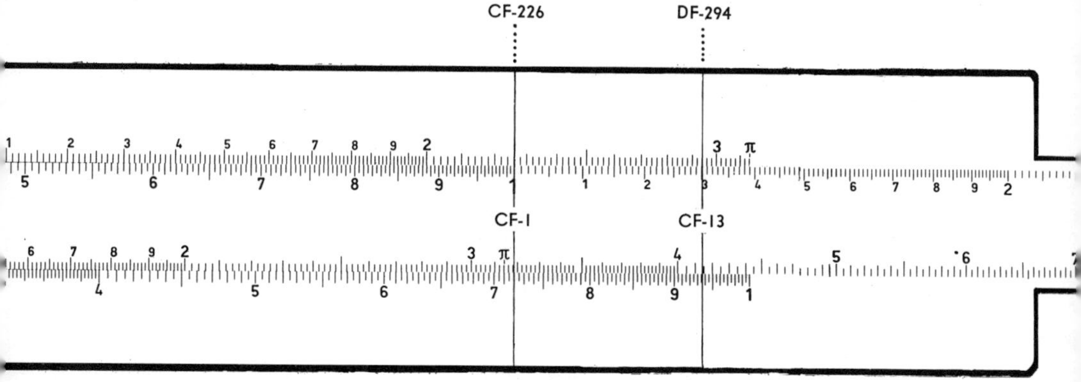

end of the slide rule. If, then, in adjusting either C-1 or C-1(right) over a D-reading, you are careful to move the slide less than half the length of the slide rule you can be assured that all possible multiplications can be made without shifting indices.

Suppose, for instance, you wanted to multiply 2.3 by 4.8. Move C-1 over D-23 and you would find that C-48 was beyond the right-hand end of the D-scale and therefore unusable. One way out would be to heave a sigh and shift indices so that C-1(right) is over D-23. This, how-

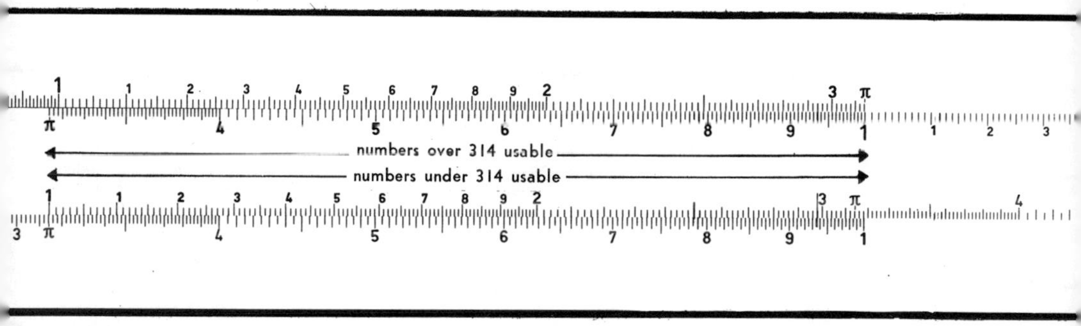

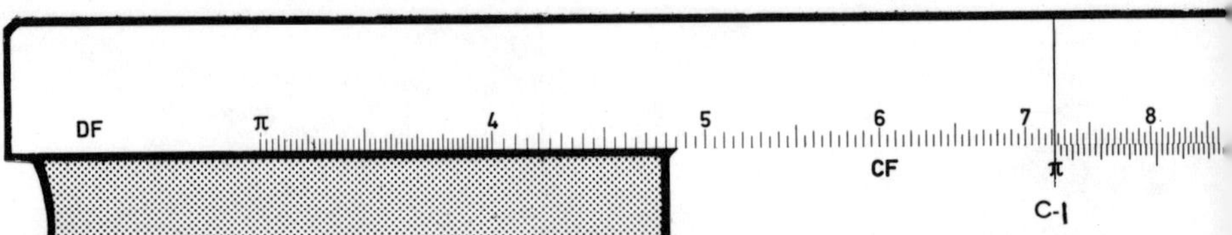

Figure 46

ever, is not necessary. Leave C-1 over D-23 and simply look over to the CF-scale. You can find CF-48 with no trouble at all. It is directly under a point between DF-110 and DF-111, somewhat closer to DF-110 (Figure 46). Call it DF-1104, therefore, and you conclude that $2.3 \times 4.8 = 11.04$.

Figure 47

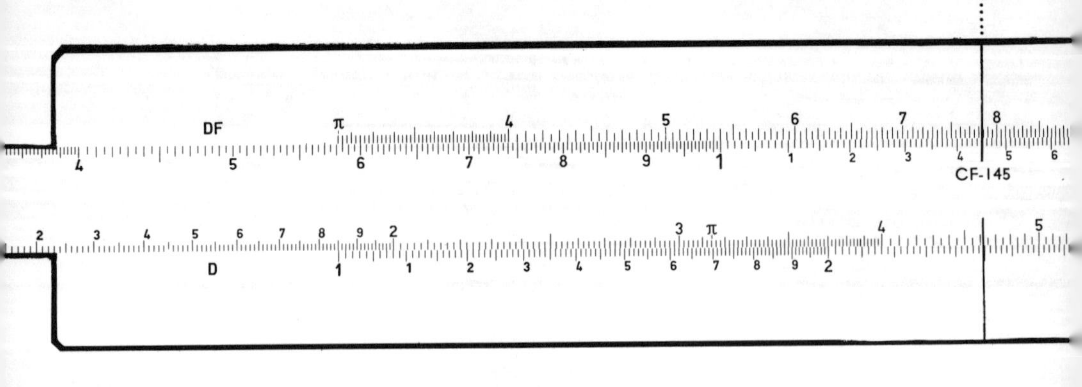

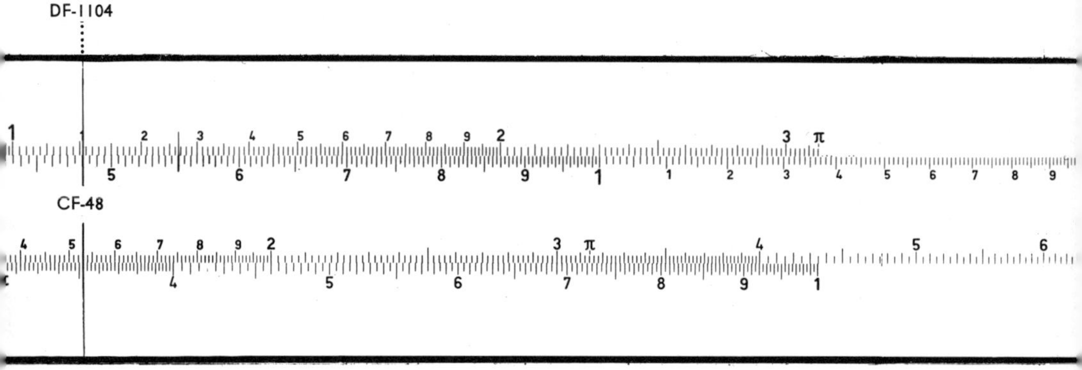

In the same way, if you wish to multiply 5.4 by 1.45 and move C-1 (right) over D-54, you will find that C-145 is beyond the left end of the slide rule. Calmly, you switch to CF-145 and find it just to the left of DF-785 (Figure 47). You judge the reading to be DF-783 and decide that 5.4 × 1.45 = 7.83.

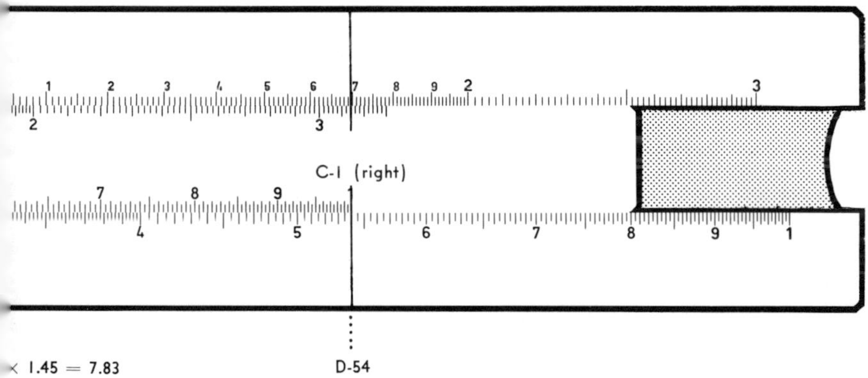

The usefulness of the folded scales is not confined only to those times when you have misjudged and used the wrong end of the C-scale. Suppose (as frequently happens) that you must multiply a whole series of numbers by a particular factor. For instance, there are just about 2.2 pounds in a kilogram and you may be given a number of weights, in kilograms, which you wish to change into pounds. Each weight in kilograms must be multiplied by 2.2 if that purpose is to be accomplished. Suppose that the kilogram weights given you are 12.3, 32.1, 46.7, and 74.3.

You move C-1 over D-22 and leave it there. That represents a general multiplication of any number by 2.2. Now you move the hairline to C-123 and find it over D-271; you move the hairline to C-321 and find it over D-706. So far, so good; 12.3 kilograms = 27.1 pounds and 32.1 kilograms = 70.6 pounds.

But C-467 and C-473 are beyond the right end of the D-scale. No bother! Don't shift indices! Just move over to the CF-scale where you will find CF-467 under DF-1027 and CF-743 under DF-1635. Therefore, 46.7 kilograms = 102.7 pounds and 74.3 kilograms = 163.5 pounds.

6

Division

Reversing Multiplication

DIVISION is the reverse of multiplication, as subtraction is the reverse of addition. When I described the addition rule, I pointed out that we could perform subtraction by running the technique for addition in reverse (see page 31). On the slide rule, we can carry out division by running the technique for multiplication in reverse.

In multiplying 2 by 3, we place C-1 over D-2 and look under C-3 to find D-6. Therefore $2 \times 3 = 6$.

In dividing 6 by 3, we place C-3 over D-6* and look under C-1 to find D-2 (see Figure 48). Therefore $6 \div 3 = 2$. The reason for this is not hard to see. The position of D-6 represents a distance of *log* 6 from the left end of the D-scale. The position of C-3 represents a dis-

* In setting a C-reading over a D-reading, where neither reading is at the end of a scale, it is best to use the hairline as a guide. To place C-3 over D-6 you would first place the hairline at D-6, then bring C-3 to the hairline. The hairline is particularly useful where the readings are not on actual markings. To place C-319 over D-413, for instance, without using the hairline, is extremely difficult, as you can assure yourself if you have a slide rule handy. With a hairline, it requires a little close attention, but is not particularly hard.

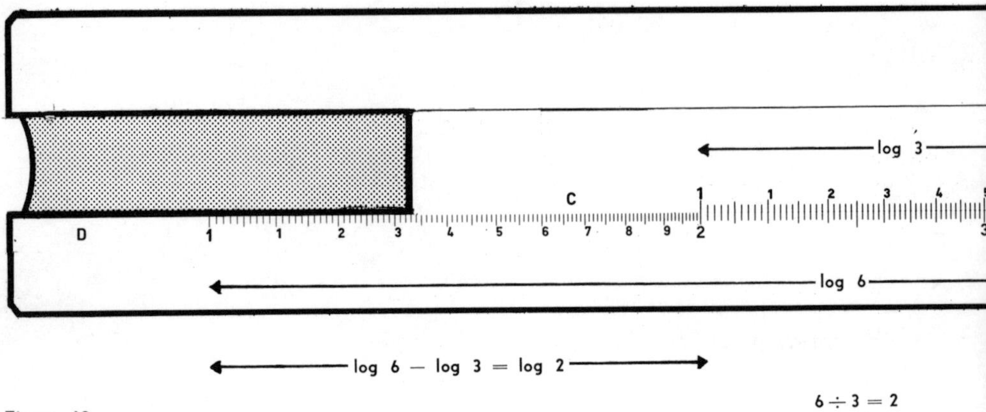

Figure 48

tance of *log* 3 from the left end of the C-scale. If we place C-3 over D-6, and follow the C-scale back to C-1, you see from Figure 48 that we are subtracting *log* 3 from the distance of *log* 6 on the D-scale. What is left of the D-scale beyond C-1 is *log* 6 − *log* 3. Since the reading under C-1 is D-2 (which is a distance of *log* 2 from the

Figure 49

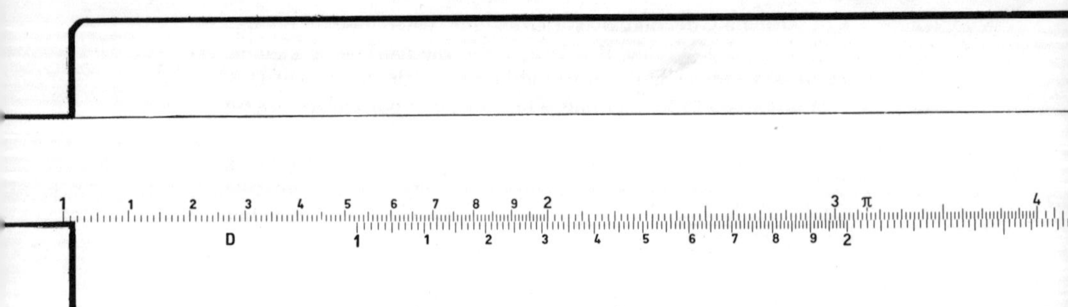

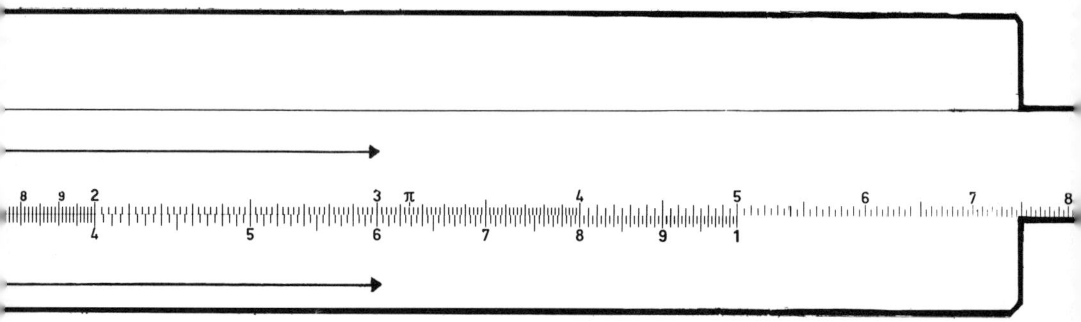

left end of the D-scale), we can say that *log* 6 − *log* 3 = *log* 2. But subtracting logarithms is equivalent to dividing antilogarithms; therefore 6 ÷ 3 = 2.

Any division can be carried through in the same fashion. If you want to divide 48.5 by 7.4, you place C-74 over D-485 and you will find that C-1 is off the scale.

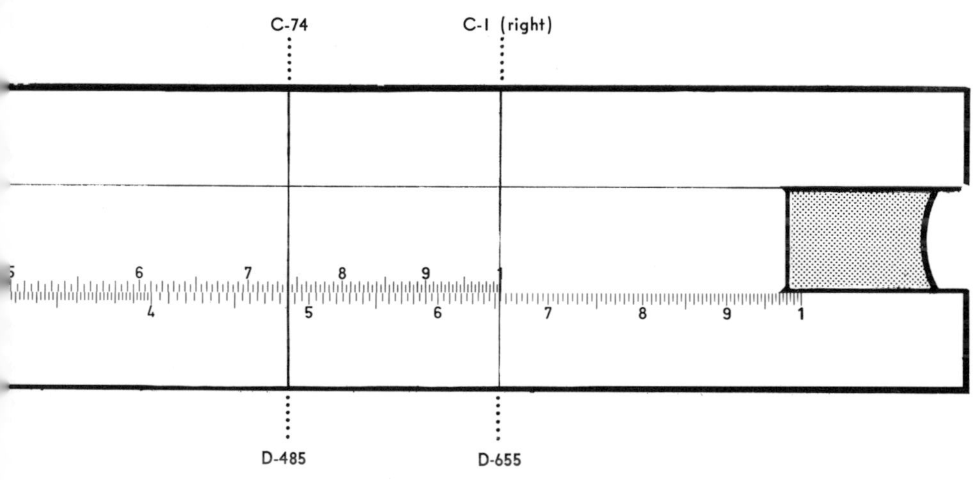

48.5 ÷ 7.4 = 6.55

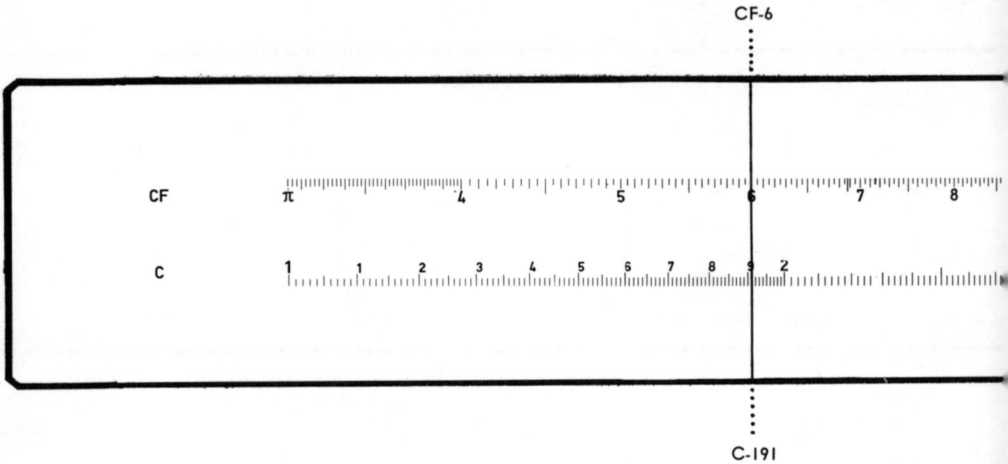

Figure 50 $6 \div \pi = 1.91$

That needn't bother you, for you need only look under C-1(right) instead and find D-655 (see Figure 49). You decide therefore that $48.5 \div 7.4 = 6.55$.

How did you determine the position of the decimal point? In the usual way. Instead of $48.5 \div 7.4$, you make use of the near-equivalent $49 \div 7$, to which the answer, as you see at once, is 7. This is not the right answer, but it is the right order of magnitude.

In dividing you place a C-reading over a D-reading and then look for either C-1 or C-1(right). No matter how you adjust the slide, either C-1 or C-1(right) will be on the D-scale so that you can always find a quotient without reference to the folded scales.

Nevertheless the folded scales do come in handy for the special case of division by π. For that purpose we can

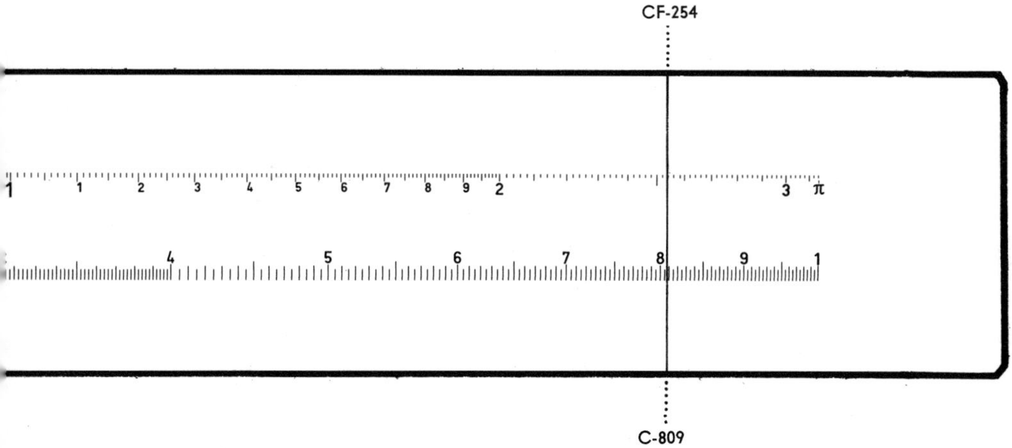

25.4 ÷ π = 8.09

make use of the CF-scale, reversing the procedure for multiplication of π (see page 113). Since the C-reading multiplied by π equals the CF-reading, the CF-reading divided by π equals the C-reading. If you want the answer to $6 \div \pi$, you have only to set the hairline on CF-6 and it will simultaneously read C-192 (see Figure 50). Since π is approximately 3 and $6/3 = 2$, you have to set your order of magnitude and can say that $6 \div \pi = 1.92$. In the same way, you can determine that $25.4 \div \pi = 8.09$, and so on.

There is one possibility of confusion in division that does not arise in multiplication. The order in which you multiply does not matter; so that 7.62×1.44 will give the same product as 1.44×7.62. Therefore it doesn't matter whether you put C-1(right) over D-762 and move the hairline to C-144, or put C-1 (right) over D-144 and move the hairline to C-762. In either case, you will end with a reading of D-110 and you will decide that 7.62 × 1.44 = 11.0.

The order in which you divide, however, does matter,

Figure 51

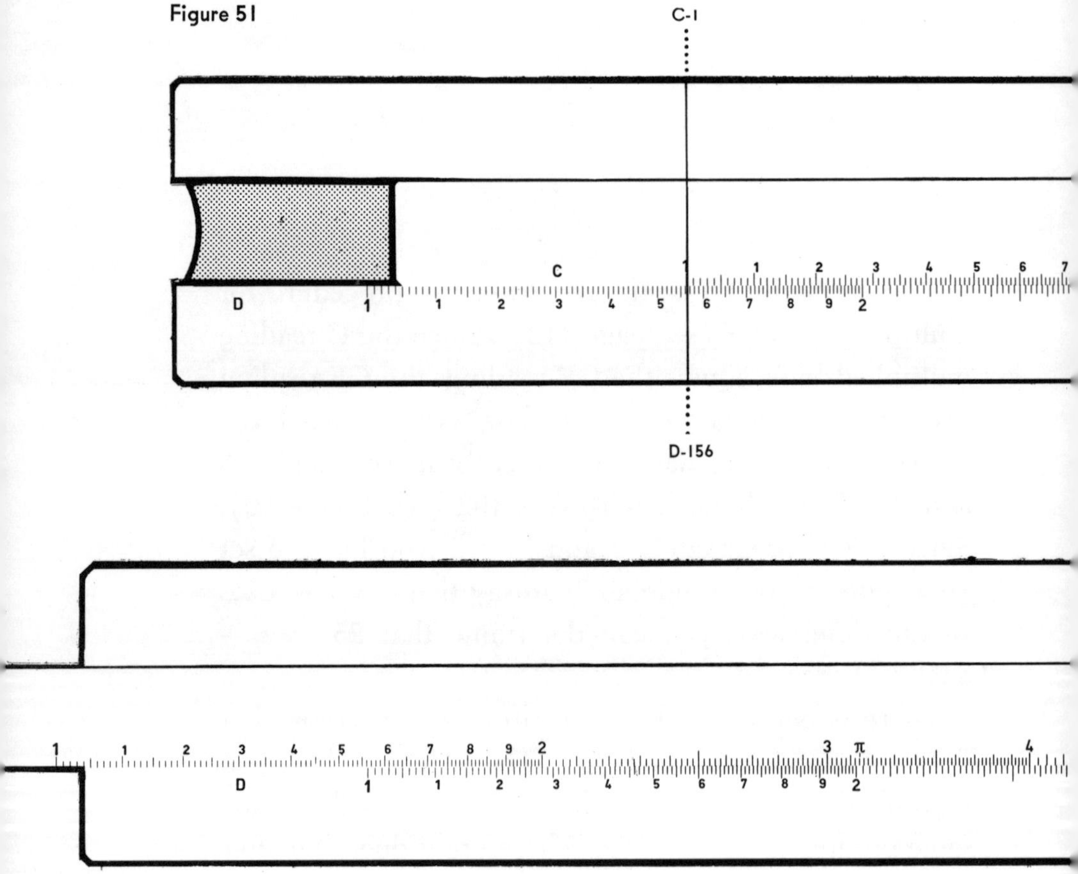

as it does when you subtract (see page 34). You can see that the quotient of 75 ÷ 48 is not the same as that of 48 ÷ 75. In the former case, you put C-48 over D-75 and find D-156 under C-1. In the latter case, you put C-75 over D-48 and find D-64 under C-1(right) (Figure 51).

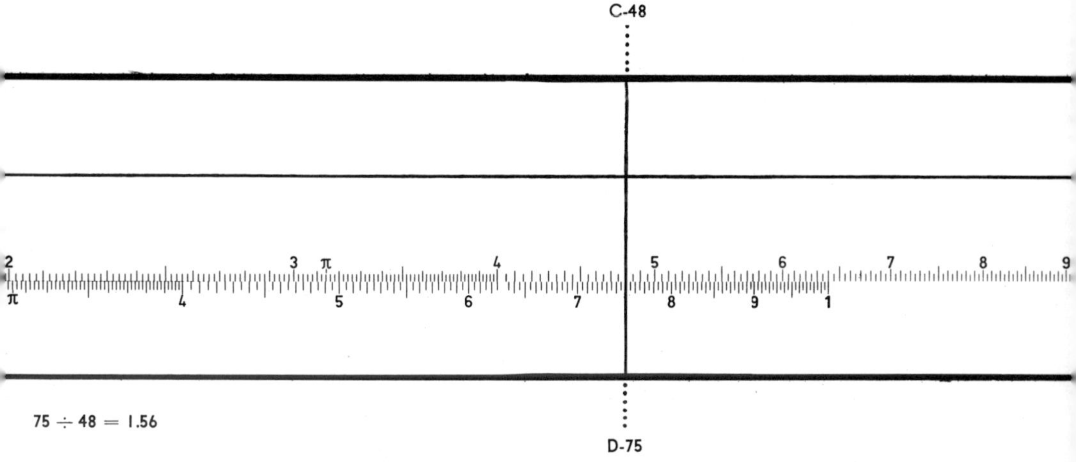

75 ÷ 48 = 1.56

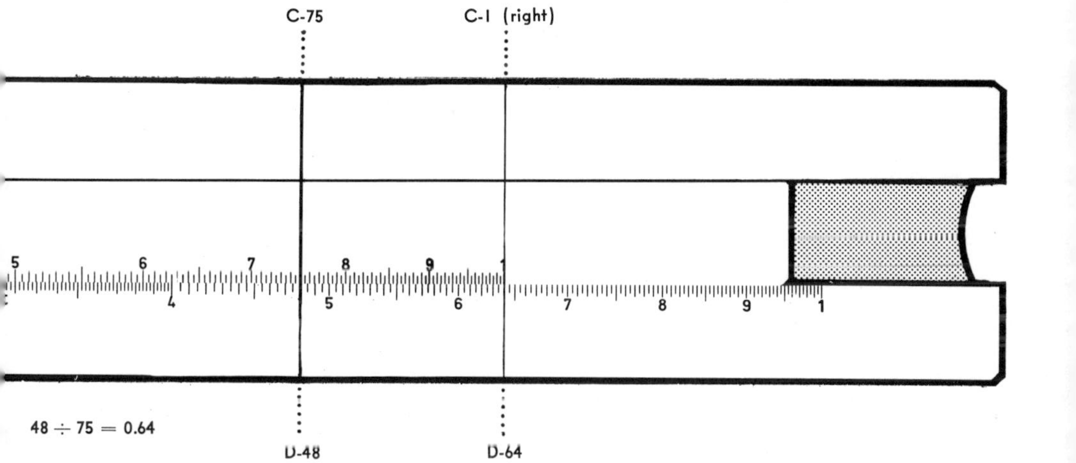

48 ÷ 75 = 0.64

Since $75 \div 48$ can be replaced by the near-equivalent $75 \div 50$, which equals 1.5, and $48 \div 75$ can similarly be replaced by $50 \div 75 = 0.67$, we see what our order of magnitude is and where to place the decimal point. We decide, then, that $75 \div 48 = 1.56$ and $48 \div 75 = 0.64$.

But how do you remember which number goes on the C-scale and which on the D-scale? When do you put C-48 over D-75 and when C-75 over D-48? Actually, the decision is a simple one.

In any division, as, for example, $a \div b$, a is the number being divided into and is the *dividend,* while b is the number doing the dividing and is the *divisor*. The rule, then, is that the divisor goes on the C-scale and the dividend on the D-scale. (This is analogous to the manner in which the subtrahend — analogous to the divisor — is placed on the slide in the addition rule. See page 39.)

In the case of $75 \div 48$, 48 is the divisor and so C-48 goes over D-75. In the case of $48 \div 75$, it is 75 that is the divisor, and it is C-75 that goes over D-48. When this rule is followed, the quotient is always found on the D-scale underneath whichever C-1 happens to be on scale.

But what if, by mistake or even by intention, the divisor is placed on the D-scale. It is not good to place matters in reverse thus, for if you follow one set rule, it will become automatic and make the use of the slide rule that much less of a conscious effort. Still, if you do reverse the rule, all is not lost.

Suppose we take the simple example $42 \div 7$ and try it first in the ordinary fashion. We place C-7 (the divisor) over D-42 (the dividend) and directly under C-1(right), we find D-6. Therefore, $42 \div 7 = 6$.

Now we'll try it "upside down." We will place C-42 over D-7 and we must now turn the procedure for finding the quotient upside down as well. Instead of finding the quotient on the D-scale under whichever C-1 is on scale, we find it on the C-scale over whichever D-1 is on scale. In this case, D-1(right) is under C-6 and again $42 \div 7 = 6$ (Figure 52).

Whichever way you carry through your division — with the divisor on the C-scale and the dividend on the D-scale, or the divisor on the D-scale and the dividend on the C-scale — you will always find the quotient on the same scale as the dividend! You must remember that.

Reciprocals

Every time you carry through a division on the C- and D-scales you are actually performing two divisions, one of which is upside-down, so to speak, compared to the other.

Let us try $23 \div 5$. We will solve it first in the ordinary way. Since 5 is the divisor, we place C-5 over D-23 and under C-1(right) we find D-46. Keeping the order of magnitude in mind, we see that $23 \div 5 = 4.6$ (Figure 53).

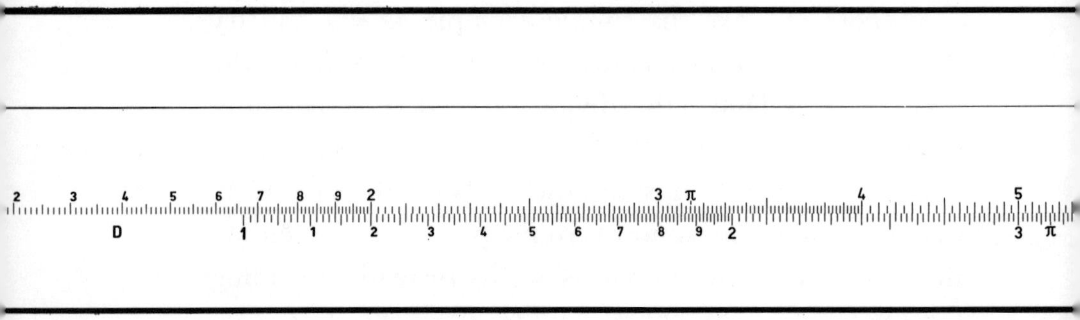

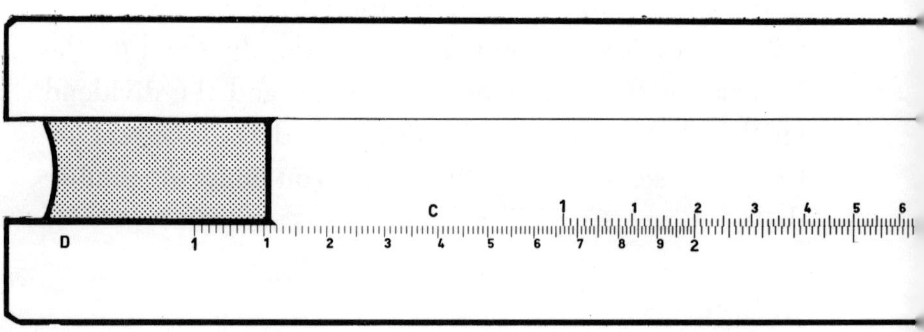

Figure 52

Take another look at the slide rule in the setting of Figure 53. That is exactly the setting we would want if we took the example $5 \div 23$ and did it upside-down,

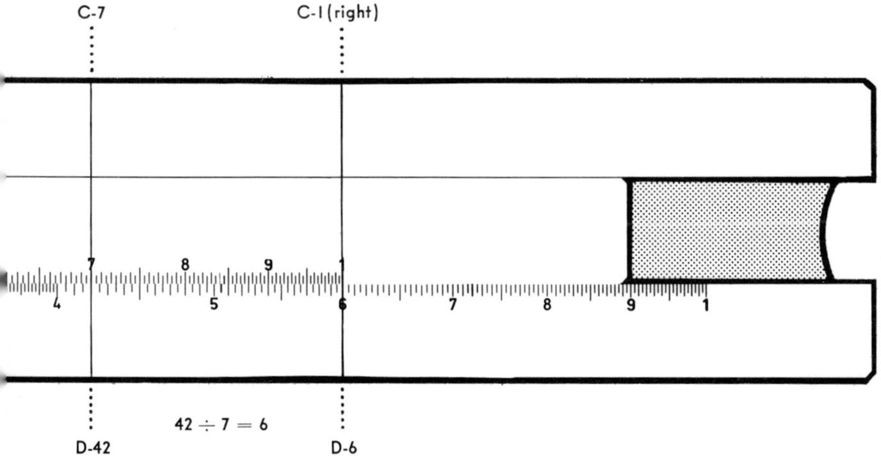

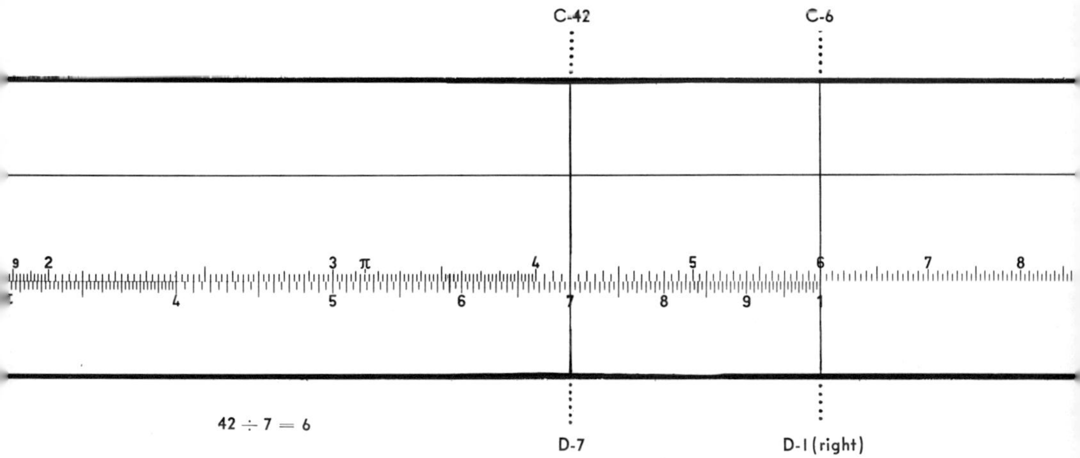

placing the divisor, 5, on the D-scale directly under the dividend, 23, on the C-scale. In that case, we will find the quotient on the C-scale, where the dividend is, and

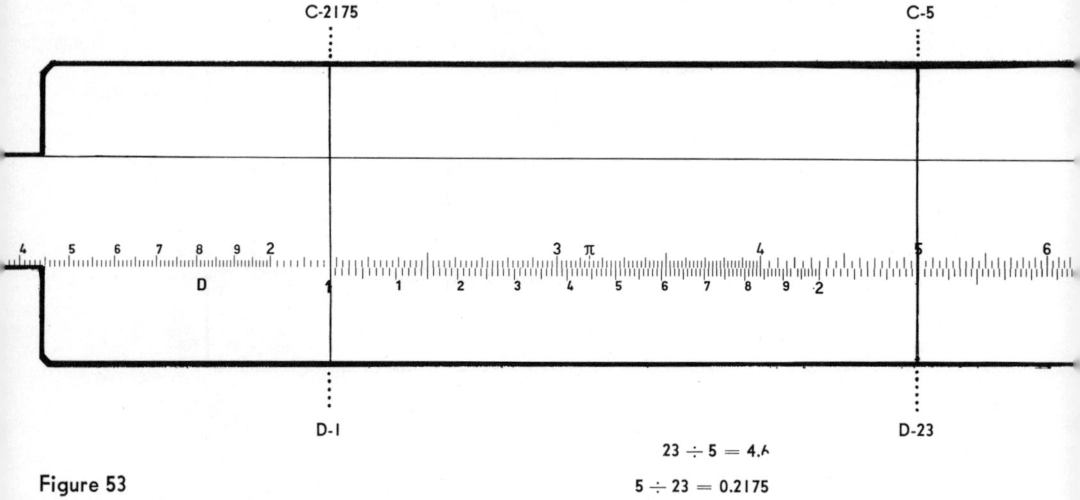

Figure 53

over D-1, we will find a reading just to the left of C-218. If we call it C-2175 and adjust the order of magnitude, we decide that $5 \div 23 = 0.2175$.

The same setting on the slide rule that gives us the quotient of any division, say $a \div b$ on the D-scale, will give us the quotient of the division $b \div a$ on the C-scale.

We can write divisions as fractions. We can write $a \div b$ as a/b and $b \div a$ as b/a. We can therefore say that the same setting of the slide rule that gives us the value of the fraction a/b on the D-scale will give us the value of the fraction b/a on the C-scale.

The fraction b/a is called the *reciprocal* of a/b and vice versa. It is often important in computations to determine reciprocals, and here is one method for doing that on the slide rule.

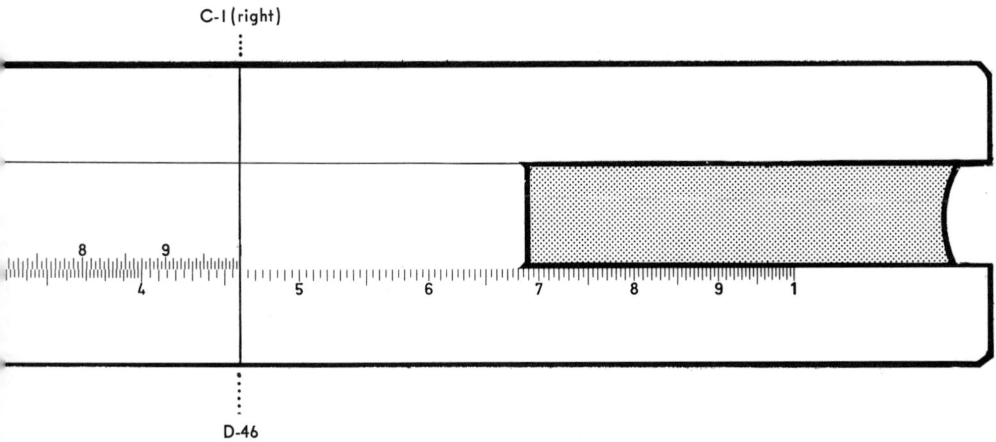

The most important reciprocals are those which involve fractions with a numerator equal to 1. The reciprocal of ¼ is 4/1, but it is customary to drop a 1 that appears in the denominator and treat such a number as consisting of the numerator alone. In other words, 4/1 is usually written simply as 4 and we can say that the reciprocal of ¼ is 4. In the same way, the reciprocal of 1/18 is 18, that of 1/29 is 29 and so on. On the other hand, the reciprocal of 17 is 1/17, that of 34 is 1/34 and so on.

Reciprocals expressed in fractional form are no problem, but what if you wish them converted to decimals. The reciprocal of 29 is 1/29, but what is the decimal form of 1/29?

Suppose we treat 1/29 as a division, as 1 ÷ 29. Since 29 is the divisor, place C-29 over D-1 and under C-1-

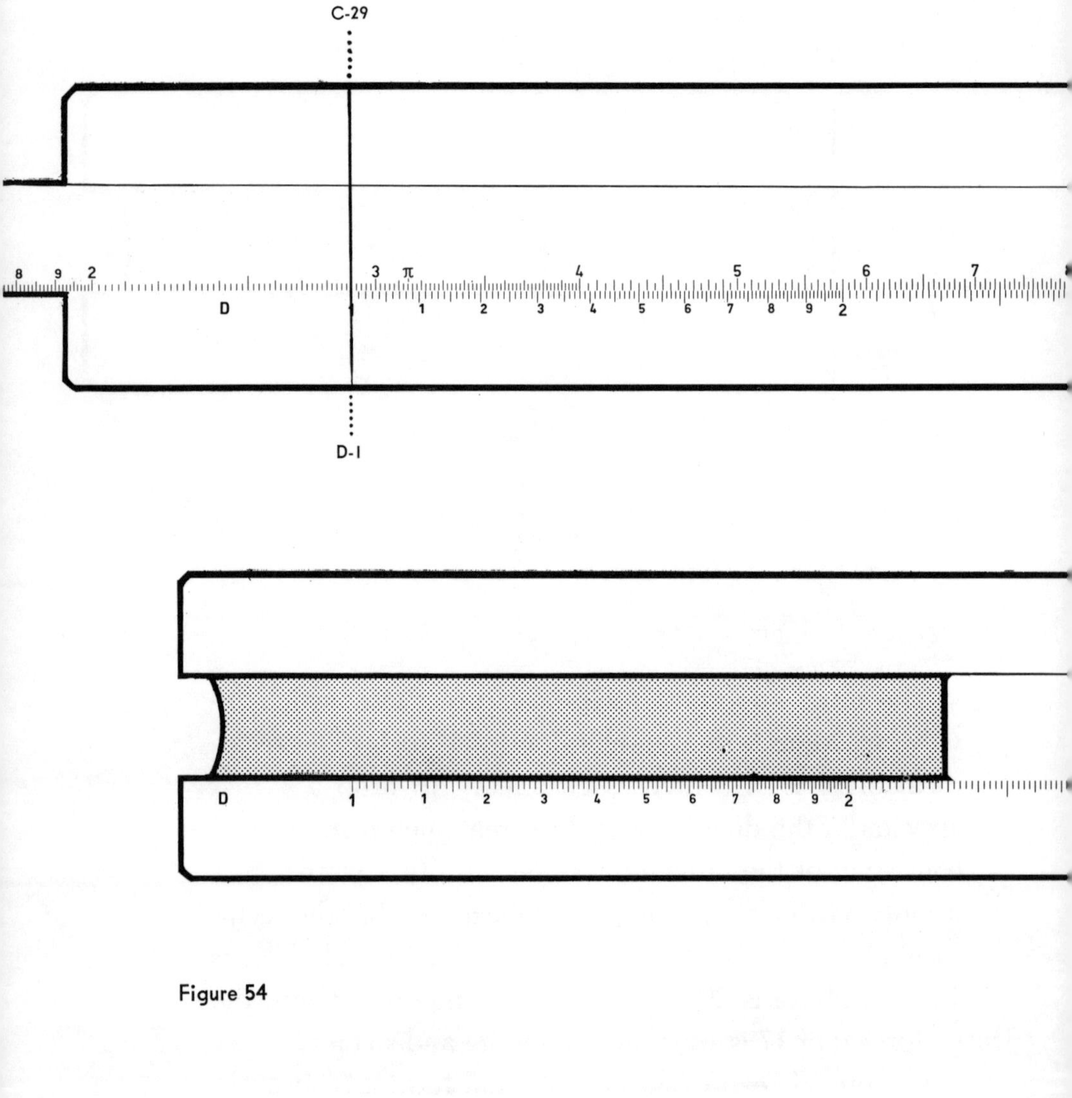

Figure 54

(right) you will find D-345 (Figure 54). Keeping the order of magnitude in mind, you decide that $\frac{1}{29} = 0.0345$. If you had placed C-29 over D-1(right), you

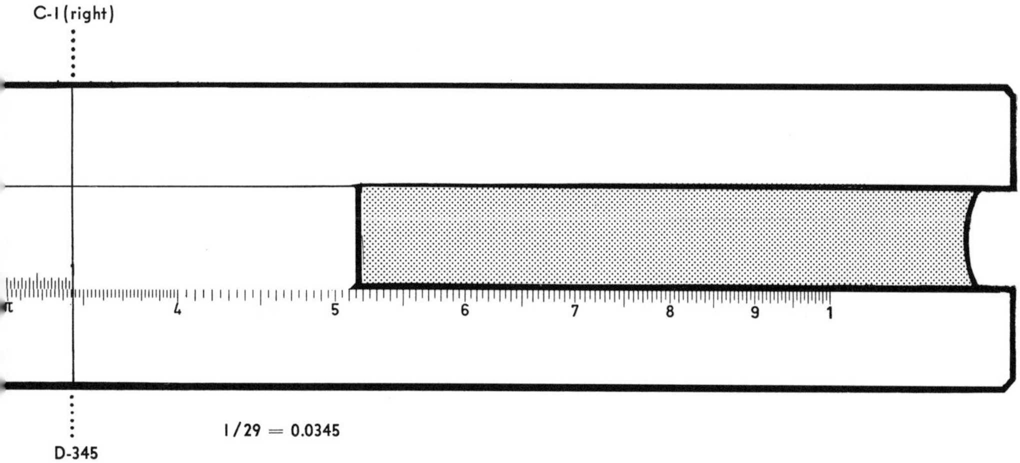

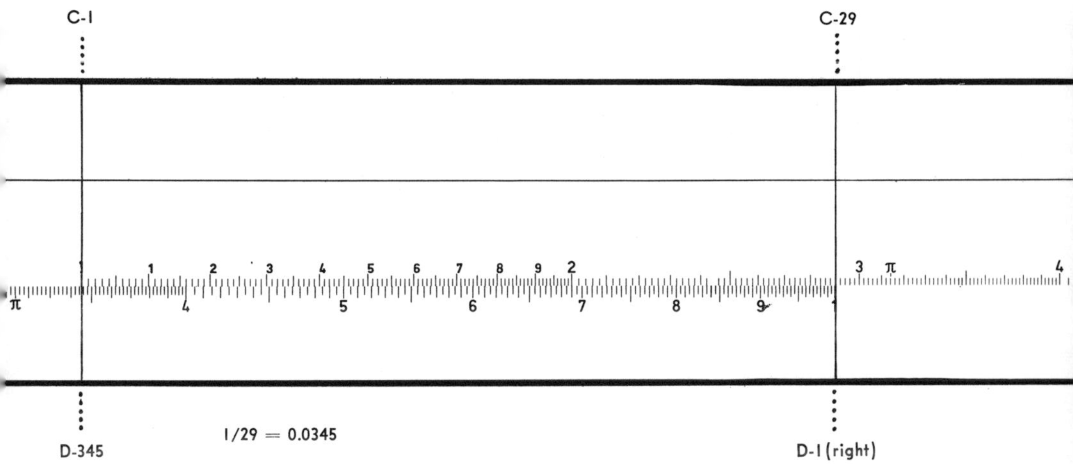

would have found, under C-1, D-345, and have obtained the same answer.

Looking at it in general, we can say that if we place

C-x over D-1, we will find D-$1/x$ under C-1(right). If we place C-x over D-1(right), we will find D-$1/x$ under C-1.

And, of course, if $1/x$ is the reciprocal of x, then we can also say that x is the reciprocal of $1/x$. If $1/29 = 0.0345$, then $1/0.0345 = 29$.

Proportion

It is possible to combine multiplication and division. Suppose you were faced with the problem: $13 \times 5.4 \div 0.83$.

Place C-1 over D-13 and move the hairline to C-54. The hairline now marks off a D-reading which is the product of 13 and 5.4, a product which you needn't look at, but which now becomes the dividend for which 0.83 is the divisor. Leaving the hairline where it is, you move C-83 directly under the hairline and under C-1(right) is D-846 (Figure 55).* Now it is necessary to place the decimal point. Consider that $13 \times 5.4 \div 0.83$ might be approximated as $10 \times 5 \div 1$, which comes to 50. Accepting that as the order of magnitude, you decide that $13 \times 5.4 \div 0.83 = 84.6$.

Some problems involving a combination of multiplication and division form a special case.

* If the problem had been $13 \times 5.4 \div 0.89$, C-1(right) then the D-reading under it would have been obscured by the glass edge of the indicator assembly — at least on my slide rule. Don't panic, however. Just move the indicator assembly and look!

Consider, for instance, two fractions, a/b and c/d, which are equal in value. We can then say that $a/b = c/d$. We can refer to such an equality of fractions as a *proportion*, and read it "a is to b as c is to d."

It may happen that you know the value of three of the four parts of a proportion and may want to find the value of the fourth. If one of the numerators is unknown, the equation becomes $a/b = x/d$. Solving for x by ordinary algebraic methods, we find that $x = ad/b$ or $a \times d \div b$. If one of the denominators is unknown, we have $a/b = c/x$ which means that $x = bc/a$ or $b \times c \div a$. In either case, we can obtain the value of x on the slide rule in the manner described just above for combining multiplication and division.

However, problems involving proportion lend themselves to particularly easy treatment on the slide rule.

Let's substitute numbers for the three known quantities and take $2/5 = 3/x$ as an example. Consider the fraction $2/5$ first. We know that its decimal equivalent is 0.4 and if we wish we can check that on the slide rule by considering the fraction to be $2 \div 5$. We place C-5 over D-2, and under C-1(right) we find D-4.

But this same quotient would be given by any other fraction that was equal to 0.4. For instance $4/10 = 0.4$, $12/30 = 0.4$, $1/2.5 = 0.4$, $6/15 = 0.4$, and so on. If, then, we tried to divide 4 by 10, or 12 by 30, or 1 by 2.5, or 6 by 15, we would end with the same setting of the slide rule with C-1(right) over D-4. Having achieved this setting

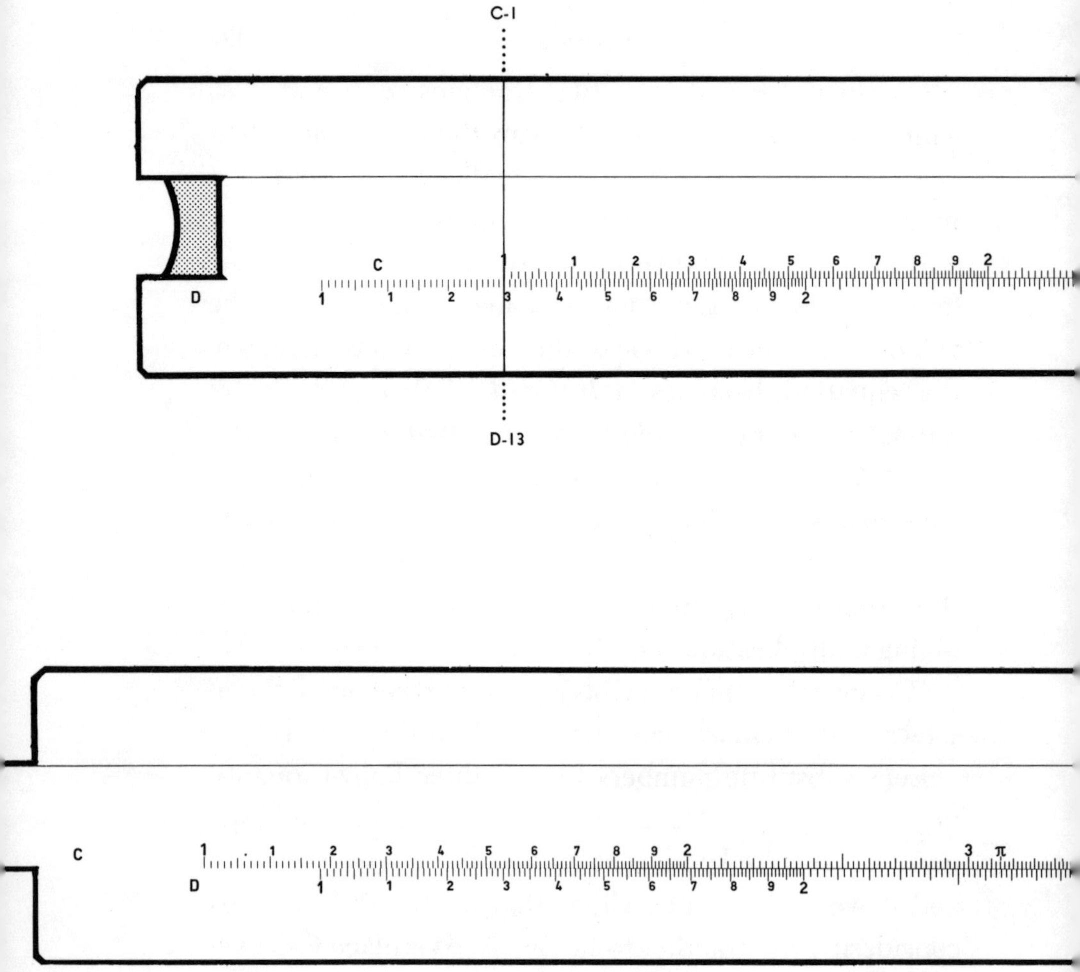

Figure 55

in the case of one such fraction, say ⅖, we ought to find, on that same setting, all the other divisions I've men-

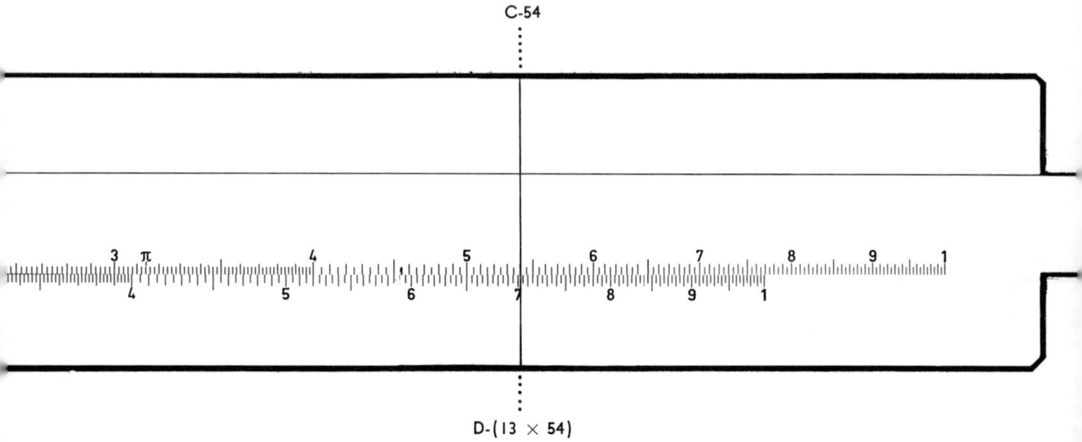

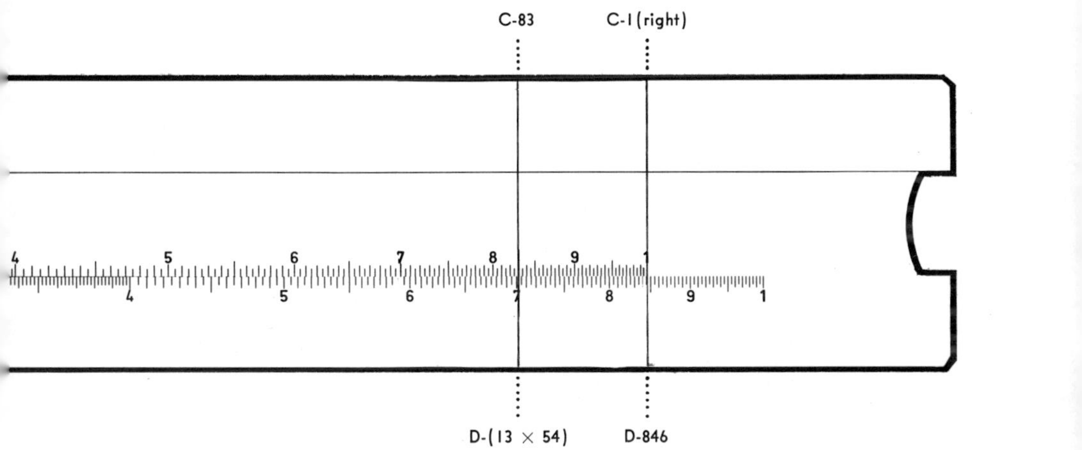

$13 \times 5.4 \div 0.83 = 84.6$

tioned and, indeed, any division at all that yields 0.4 as the quotient. Sure enough, if you'll look at Figure 56,

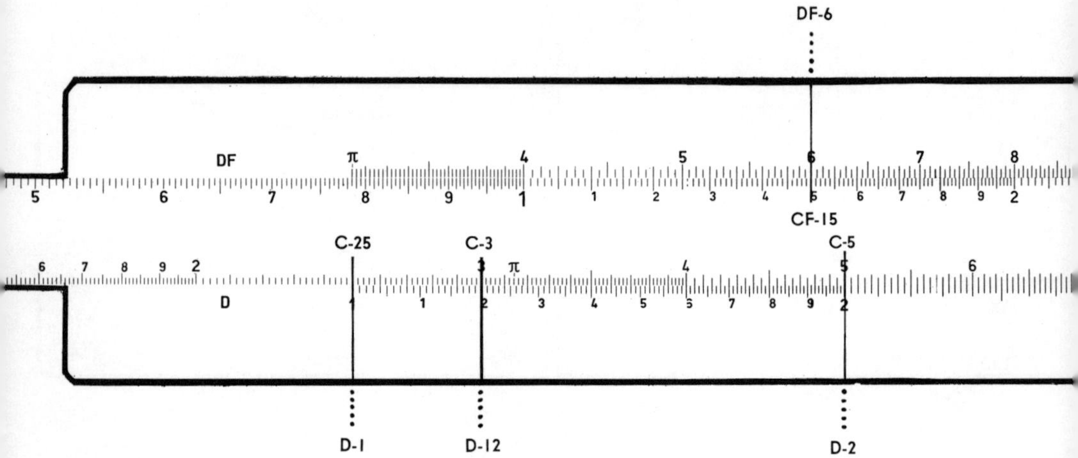

Figure 56

you will find that just as C-5 is over D-2, you have C-1 over D-4, C-3 over D-12, C-25 over D-1, situations which are equivalent to $\frac{2}{5}$, $\frac{4}{10}$, $\frac{12}{30}$, and $\frac{1}{2.5}$.

To be sure, we can't find $\frac{6}{15}$ on the C- and D-scales because C-15 is far to the left of the end of the D-scale. However, if we switch to the CF- and DF-scales (see page 110), we will find CF-15 under DF-6.

In short, once you set up a given fraction on the slide rule, with the denominator on the C-scale directly over the numerator on the D-scale, then all other fractions of equal value are similarly given (denominator on C-scale, numerator on D-scale) on other parts of the slide rule. If, on the other hand, you wish to place the fraction with the numerator on the C-scale over the denominator on

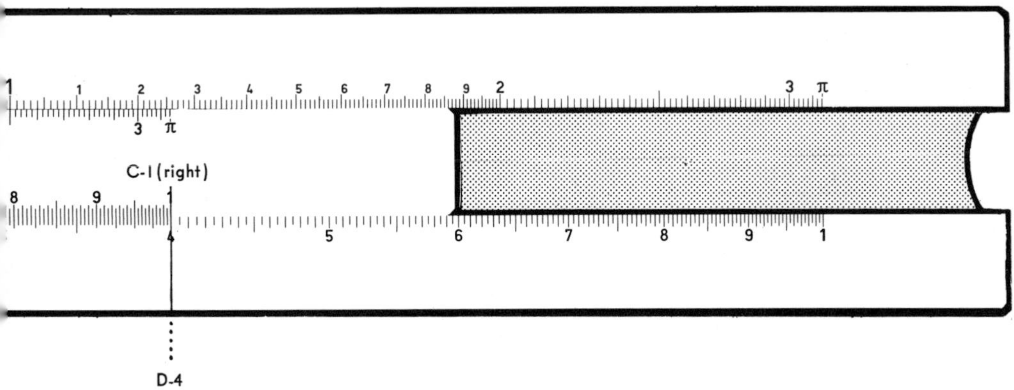

the D-scale, then you will find other fractions with numerator and denominator similarly placed.

In a proportion, two fractions are set equal to each other. If one fraction is set upon the C- and D-scales in the manner described, the other can also be found.

If we go back to $2/5 = 3/x$ and place C-5 over D-2, then we should expect to find C-x over D-3. We move the hairline to D-3 and immediately above it is C-75. Since 2 and 5 in the first fraction are of the same order of magnitude, 3 and x should also be of the same order of magnitude. We therefore place the decimal point so as to say $2/5 = 3/7.5$.

In the same way, if we are faced with the proportion $5.12/277 = x/34.3$, we place C-277 over D-5.12 and move

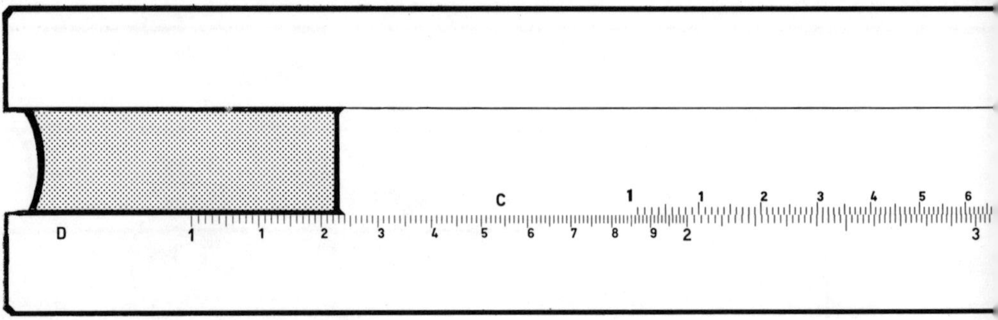

Figure 57

the hairline to C-343, finding that to be over D-633 (see Figure 57). Since 5.12 in the first fraction is two orders of magnitude smaller than 277, we expect x to be two orders of magnitude smaller than 34.3. We decide, then, that $5.12/277 = 0.633/34.3$.

Inverse Scales

The key respect in which slide rule division differs from slide rule multiplication is that in multiplication, we set C-1 and find the answer under the multiplier, while in division, we set the divisor and find the answer under C-1.

This is to be expected, since division is the inverse of multiplication. When I was explaining the workings of

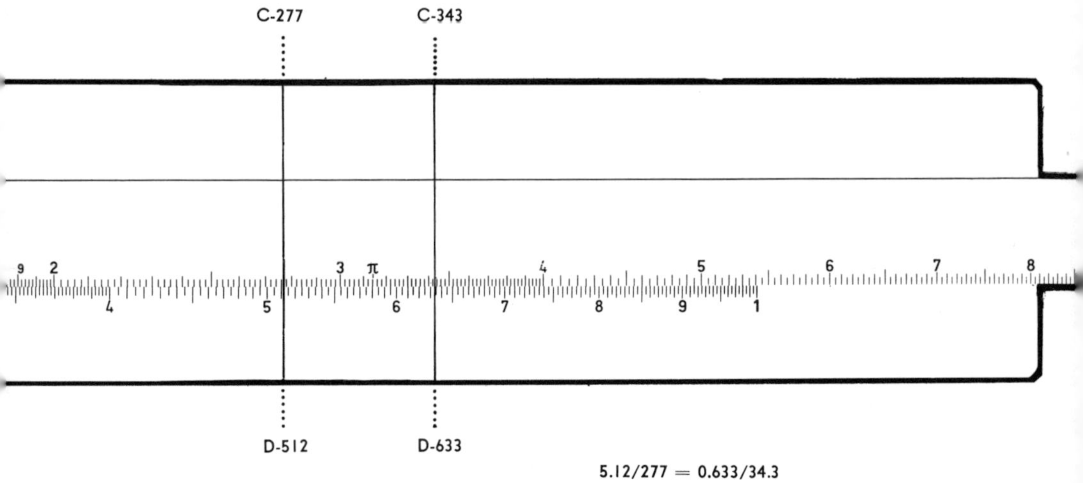

the addition rule, there was this same inverse effect in connection with subtraction as compared to addition (see page 33).

Once you are thoroughly used to working the slide rule in multiplication and division, you will find no difficulty in setting C-1 in multiplication but not in division. Nevertheless, there is some usefulness to preparing a situation in which you can set C-1 for both division and multiplication.

To make this possible, we must do for division what we did for subtraction in the addition rule. We must prepare an *inverse scale,* one that "runs backward." An inverse logarithmic scale would have its numbers crowding to the left, in mirror-image to a normal scale.

Such an inverse scale is placed in my slide rule imme-

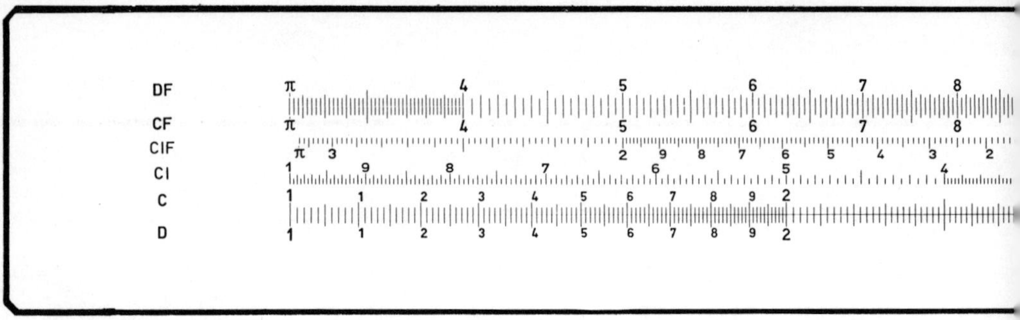

Figure 58

diately above the C-scale. It is a "C-inverse" or *CI-scale*. Above the CI-scale and immediately under the CF-scale is another inverted scale, a folded one. It is the mirror image of the CF-scale, in fact, and is, therefore the CIF-scale (Figure 58.)

To distinguish these inverse scales from ordinary ones, and to make sure that the casual slide rule user notices something is different, the CI- and CIF-scales are usually printed in red, whereas all the other scales I have discussed are printed in black.

The CF-scale can be used in division, just as an in-

Figure 59

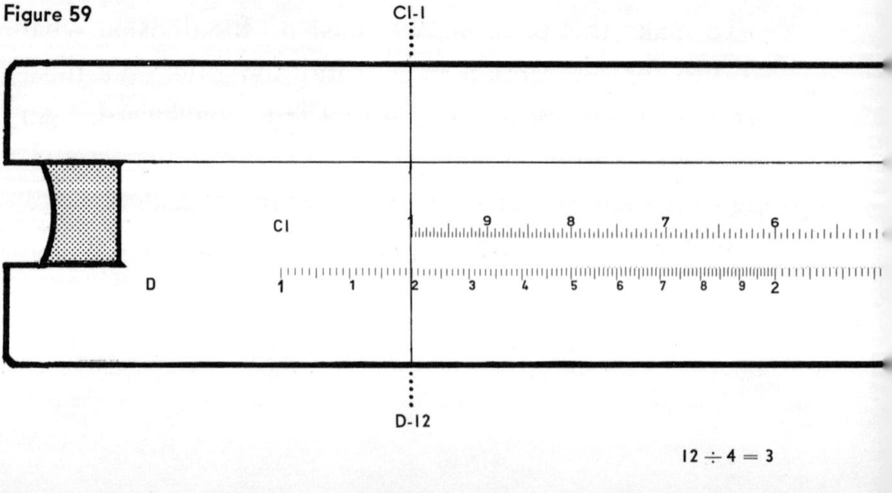

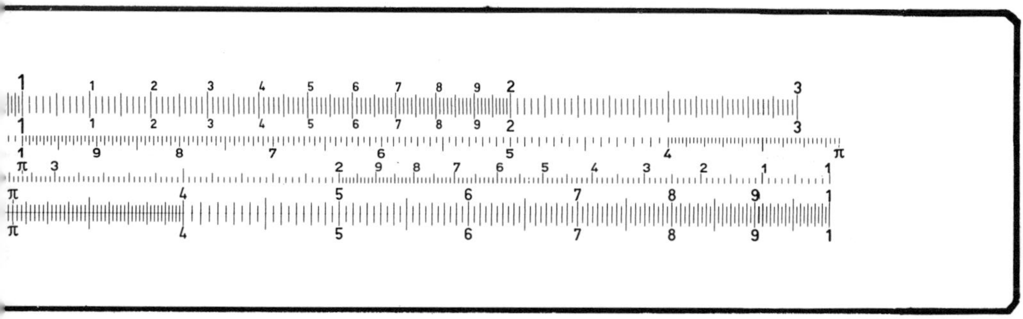

verted scale was used in subtraction on the addition rule (see page 37).

Consider the problem 12 ÷ 4. If this is carried through in the ordinary manner, using the C- and D-scale, C-4 is brought over D-12 and under C-1(right) is found D-3, showing that 12 ÷ 4 = 3.

Suppose, however, we use the CI-scale along with the D-scale. Again, we will deal with the problem 12 ÷ 4. This time we place the CI-1 over D-12 and move the hairline to CI-4. Under it is D-3, the quotient (Figure 59). Suppose we had tried to solve 45 ÷ 9 in this way.

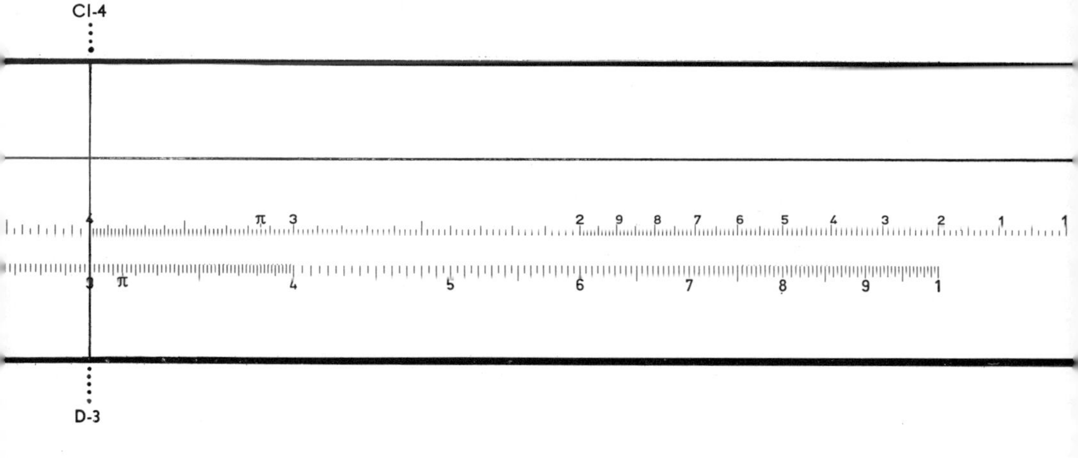

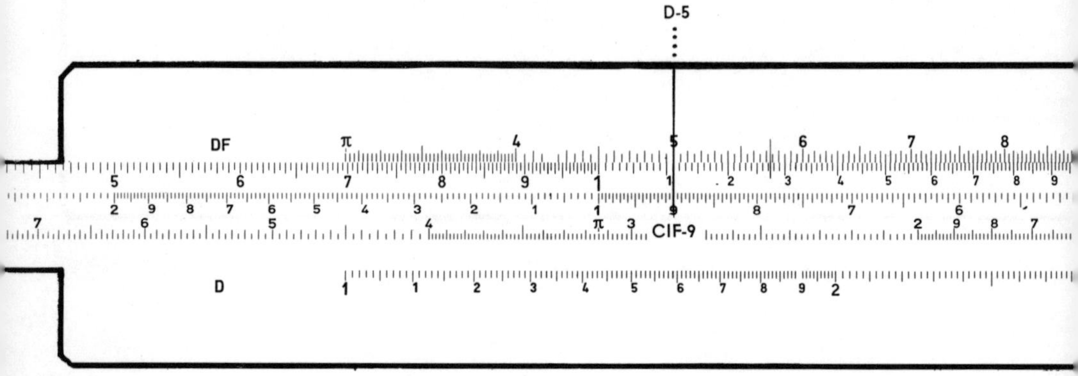

Figure 60

We place CI-1(right) over D-45, but find we cannot move the hairline to CI-9, for that is far off the D-scale. We switch to the CIF-scale, therefore. Placing the hairline on CIF-9, we find it under DF-5, so that $45 \div 9 = 5$ (Figure 60). (Had we placed CI-1 over D-45, we would not have had to switch to the folded scales.)

By using the inverse scales, in other words, we evolved a routine in which we set the end of a scale (C-1 in multiplication and CI-1 in division) in carrying through both multiplication and division.

The inverse scales can be used for another purpose, too. To see what that is, let us consider some reading on the C-scale, which we can call x, and the reading immediately above it on the CI-scale, which we will call y.

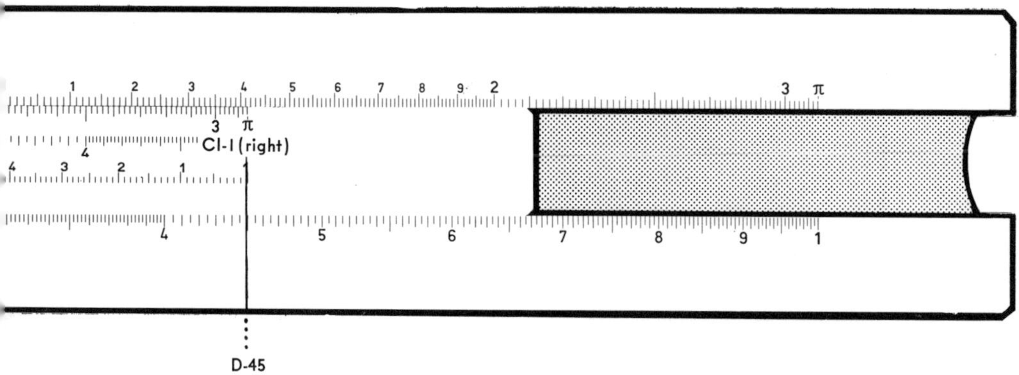

45 ÷ 9 = 5

The distance of x from C-1 on the left represents the logarithm of x. The distance of y from CI-1(right) — remember the CI-scale is a "backward" one — represents the logarithm of y. Therefore, the entire stretch from C-1 to CI-1(right) is equal to $\log x + \log y$ (see Figure 61).

But the distance represented by the entire stretch from C-1 to CI-1(right) is equal to $\log 10$. We can therefore say that $\log x + \log y = \log 10$. If we switch from logarithms to antilogarithms we say that $x \times y = 10$, or $x = 10/y$.

We can, of course, adjust the decimal point to suit ourselves. For instance, we can move the decimal point of y in such a way as to divide its value by ten. In that case, in order to keep the value of $10/y$ unchanged, we

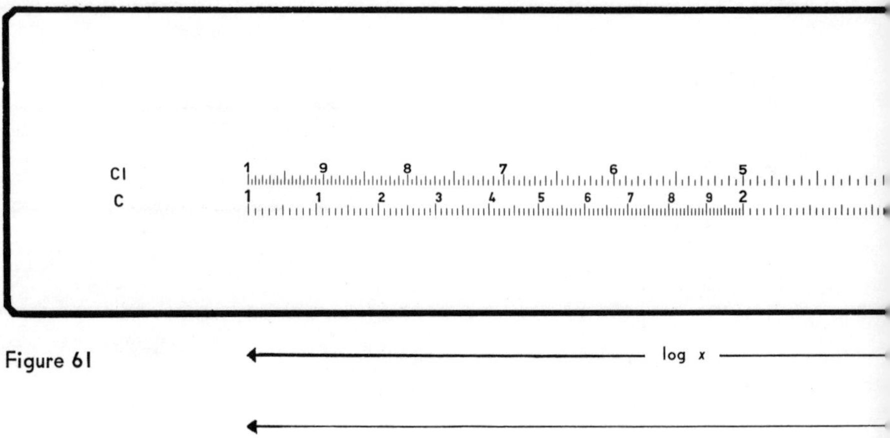

Figure 61

must also divide the numerator of the fraction by ten. We end by saying that $x = 1/y$. (For that matter we can also, with full justification, say that $x = 100/y$ or $x = 0.001/y$, but it is most convenient to say that $x = 1/y$.)

If $x = 1/y$, then x is the reciprocal of y and vice versa. It follows then that when the hairline crosses the slide, it marks off simultaneous readings on the C-scale and CI-scale that are reciprocals of each other. You can see for instance that CI-2 is directly over C-5. This tells us (if we make sure to keep the decimal point in the proper position) that the reciprocal of 2 is 0.5 and vice versa, since $2 = 1/0.5$ and $0.5 = \frac{1}{2}$. The same reading tells us

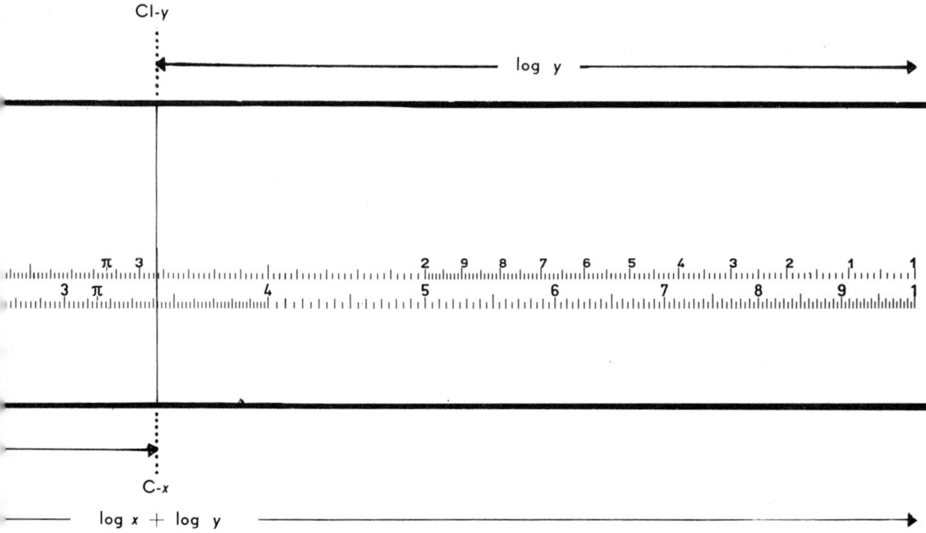

that the reciprocal of 20 is 0.05, that the reciprocal of 5 is 0.2, that the reciprocal of 0.005 is 200, and so on, since $\frac{1}{20} = 0.05$, $\frac{1}{5} = 0.2$, and $1/0.005 = 200$.

For none of these reciprocals is the slide rule really needed; we can work it out in our heads without trouble. The principle is established, however.

Suppose, now, you wanted the reciprocal of 485. First let's place the decimal point. Fractionally, the reciprocal of 485 is $\frac{1}{485}$ which is slightly larger than $\frac{1}{500}$. Since $\frac{1}{500} = 0.002$, we expect the reciprocal of 485 to be slightly larger than 0.002.

Now place the hairline on C-485 and you will find it

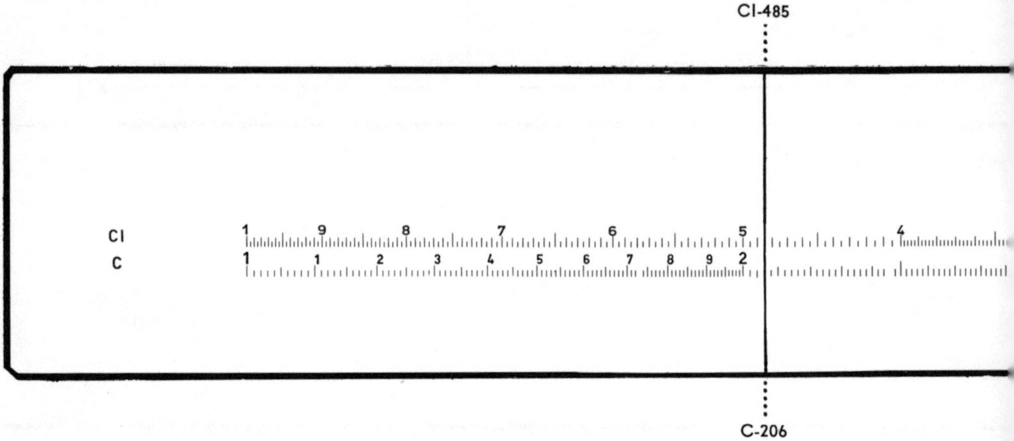

Figure 62

to be marking CI-206 at the same time (Figure 62). The reciprocal of 485 is therefore 0.00206. (If you had placed the hairline on CI-485, you would have found it to be marking off C-206. The same is true of the CF-scale and the CIF-scale taken in combination. A reading of CF-485 corresponds to one of CIF-206, and a reading of CF-206 corresponds to one of CIF-485.)

Reciprocals can also be found on the C- and D-scales by the method I described earlier (see page 130). The use of the inverse scales, however, is the simpler for such a purpose.

I will describe one other adventure in reciprocals. As you know, for any C-reading, say C-x, the corresponding CF-reading is πx (see page 113). In other words, corresponding to C-2 is CF-628, so that $2\pi = 6.28$.

Since the CIF-readings are the reciprocals of the corresponding CF-readings, for the setting C-x the corre-

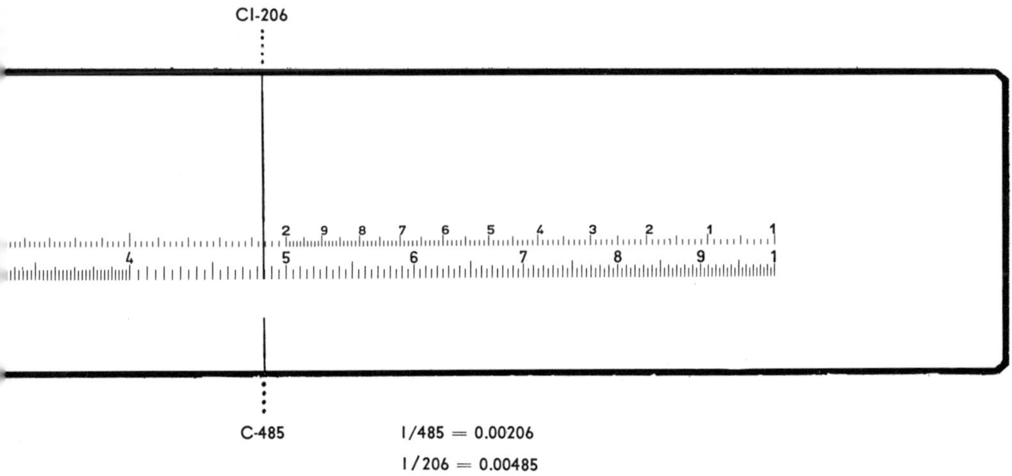

sponding reading on the CIF-scale is equivalent to $\frac{1}{\pi x}$.

Let's take a specific example. Corresponding to C-2, we have CIF-1591 (Figure 63). This means that $\frac{1}{2\pi} =$ 0.1591. Placing the decimal point is not difficult, if we remember that π is close to 3. In that case $\frac{1}{2\pi}$ is approximately $\frac{1}{6}$ or 0.167. That sets the order of magnitude at once.

Similarly, if you want the value of $\frac{1}{8.2\pi}$, place the hairline of C-82 and find the simultaneous reading CIF-388. Since 8.2π is close to 25, we can say that $\frac{1}{8.2\pi}$ is close to $\frac{1}{25}$ or 0.04. That fixes the order of magnitude and we conclude that $\frac{1}{8.2\pi} = 0.0388$.

7

Powers

The Other Side

So FAR, I have described seven scales, one under the other, on the slide rule. On the upper part of the body is the DF-scale. On the slide are the CF-scale, CIF-scale, CI-scale, and C-scale. On the lower part of the body are the D-scale and the L-scale.

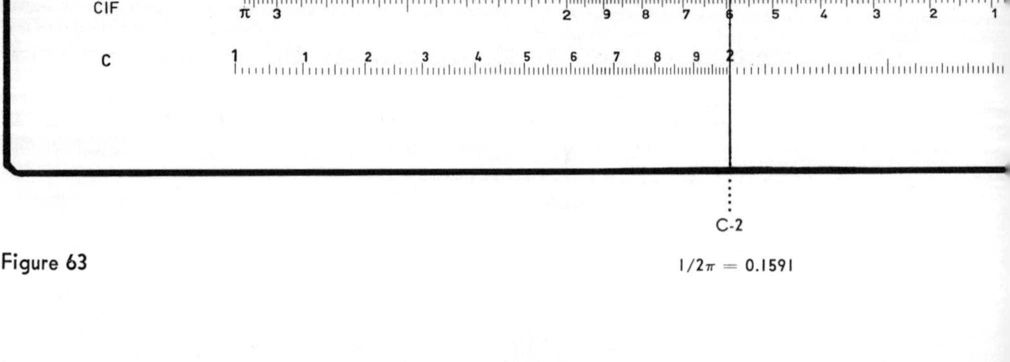

Figure 63

$1/2\pi = 0.1591$

There is room for more and, indeed, one slide rule I use has several scales in addition to those I have listed. These additional scales, however, involve operations which will not be taken up in this book, and therefore, these scales will not be discussed.

Even so there remains the other side of the slide rule which is available for other scales, some of which can be used for operations we will consider in this book.

Let's take up the other side of the slide rule, then. We can begin by pointing out that there are two ways of reversing the slide rule. It can be rotated horizontally in a long swing, or vertically in a short flip.

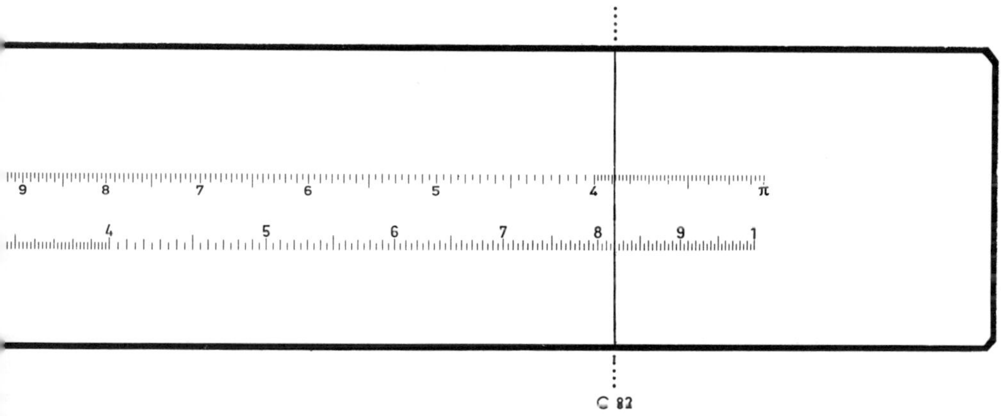

$1/8.2\pi = 0.0388$

Obviously, the short flip is easier and it is the method of rotation used. The proof of that is, that if you rotate a slide rule horizontally, you will find that the markings on the reverse side are upside-down. If you flip it vertically, the markings on the reverse side are right-side up. You are clearly expected to perform the latter movement.

In flipping the slide rule, however, we reverse the position of the top and bottom of the body. What was the upper body in front becomes the lower body in back and vice versa. You can check this by placing one finger on the upper body and keeping it there while you flip.

The scales on the other side of my slide rule are, with one exception, different from those on the front. That one exception is the D-scale, which is present on the reverse side in the same D-position (see page 73) as in front.

To distinguish this D-scale from the one we have been using previously, we will call it the *D(back)-scale.* Notice that the D(back)-scale is not on the same part of the body as the D-scale. They both seem to be on the lower body if we look at each separately, but in order to look first at one, then at the other, we must flip the slide rule. If you place your thumb on the D-scale and your forefinger on the D(back)-scale, you will see that the two are on different parts of the body.

If the slide rule is properly adjusted; if one half of the

body is fastened directly over the other and if the two glass windows of the indicator assembly are properly positioned, so that the hairline of one is directly opposite the hairline of the other, then the two D-scales will be coordinated. If the hairline gives a reading of D-x, it will be found to give a reading of D(back)-x, when the slide rule is flipped.

The D(back)-scale acts as a connecting link between the two sides of the slide rule. Most often, the answer obtained in an ordinary multiplication or a division will be found on the D-scale. If there is then anything you wish to do to that answer which will involve scales found only on the other side, you need only flip your slide rule. There is the answer found on the D-scale, properly marked off on the D(back)-scale, without any necessity on your part to make a new setting. And you may then continue.

Under the D(back)-scale, for instance (on my slide rule), is a *DI-scale,* which is the red-marked inverse of the D-scale, just as the CI-scale is the red-marked inverse of the C-scale (see page 144).

The D(back)- and DI-scales can be used, together, to find reciprocals, just as the C- and CI-scales or the CF- and CIF-scales can be so used (see page 148).

There are two advantages, though, to using the D- and DI-scales. In the first place, if you are working on the other side of the slide rule, you can find your reciprocals right there.

The less obvious one (for my slide rule) is that the DI-scale is the only inverse scale on the body and not on the slide. This means that reciprocals can be found regardless of the position of the slide.

Suppose, for instance, you wanted the solution to a problem like $\frac{1}{6.2 \times 1.8}$. To obtain the product of 6.2×1.8 is simple, but once that is obtained what you want is the reciprocal.

Let us begin. We place C-1(right) over D-62, move the hairline to C-18 and find that to be over D-1116. What we need, now, is the reciprocal of 1116. This means, if we are to use the C- and CI- scales, or the CF- and CIF-scales, that we must move the hairline to 1116 on one of those four scales.

Yet that is not necessary. If we flip the slide rule, the D-reading is duplicated in the D(back)-reading and the DI-reading is the reciprocal of that. We need move neither slide nor hairline.

In short, let's try to solve the problem $\frac{1}{6.2 \times 1.8}$ again. Place C-1(right) over D-62, move the hairline to C-18, flip the slide rule and note that the hairline marks DI-896 (Figure 64). Next we will check the decimal point. Since $\frac{1}{6.2 \times 1.8}$ can be approximated as $\frac{1}{6 \times 2}$, the answer must be about $\frac{1}{12}$ or 0.083. Hence, $\frac{1}{6.2 \times 1.8} =$ 0.0896.

The A- and B-scales

On the reverse of the slide rule in the A- and B-positions (see page 73) are two scales actually named for those positions, the *A-scale* and the *B-scale* (Figure 65).

The A- and B-scales are a new variety. Except for the L-scale, all the scales I have discussed so far have really been members of a single family represented most clearly by the C-, D-, and D(back)-scales. The CF- and DF-scales differ from these only in that they start at the middle; the CI- and DI-scales only in that they run backward; the CIF-scale only in that it starts at the middle and runs backward also. All these, however, have the same system of primaries, secondaries, and tertiaries.

Not so the A- and B-scales. They begin at the left at 1 and pass through the units, reaching 1 again in the middle. There is then room for an exact repetition, reaching 1 still again at the right end. It is as though the C- and D-scales have been compressed to half their length so that two of them will fit into the usual slide rule length.

We can refer to the two identical halves of these scales, as the *A(left)-scale* and *A(right)-scale*, and, of course, the *B(left)-scale* and *B(right)-scale*.

Let's stop to consider what this means. You may remember that I described the construction of the C- and D-scales by lining up their markings as antiloga-

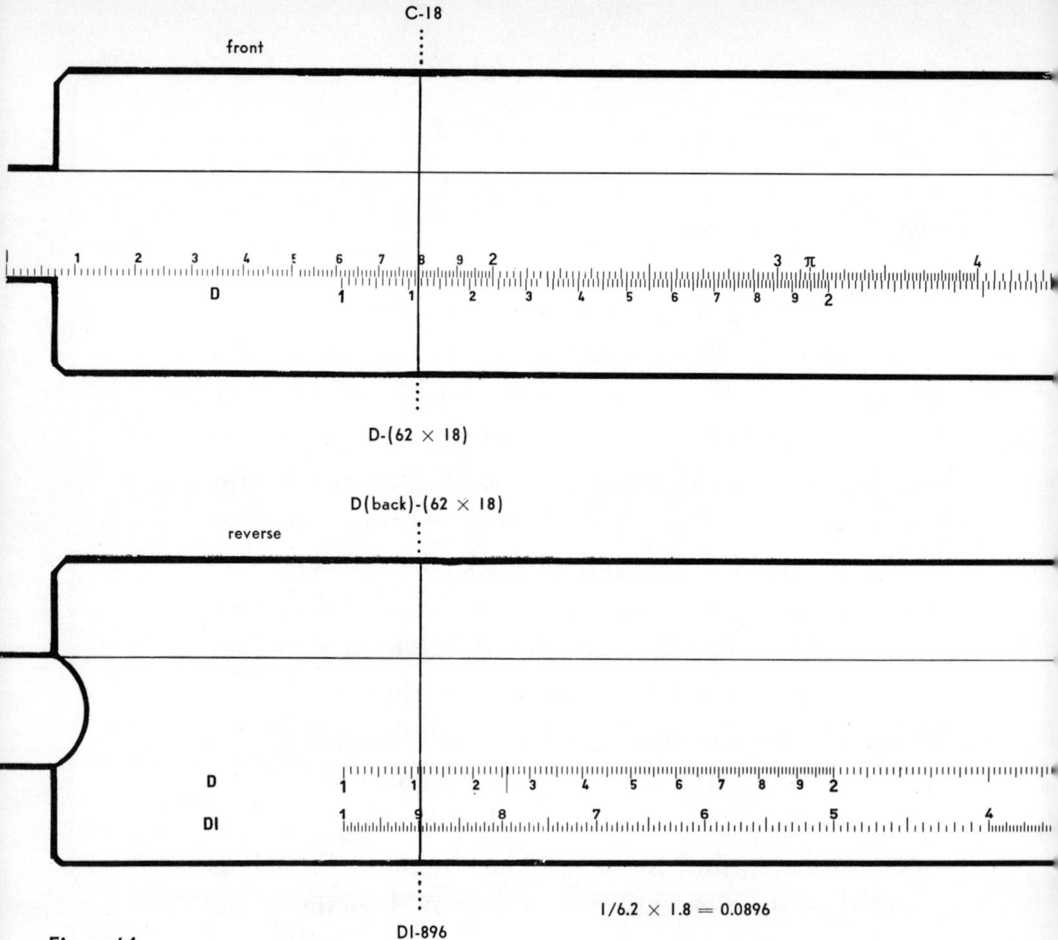

Figure 64

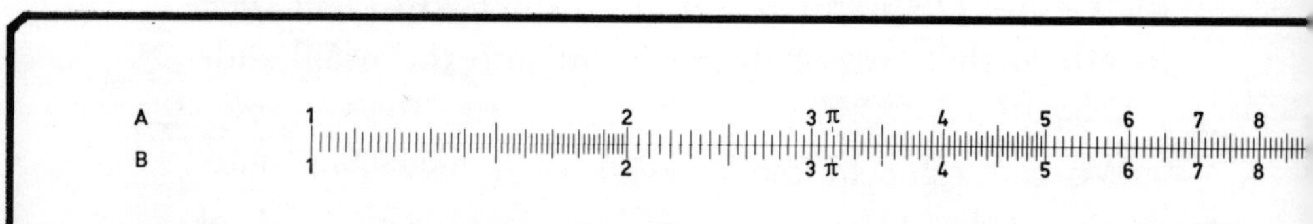

Figure 65

C-I (right)

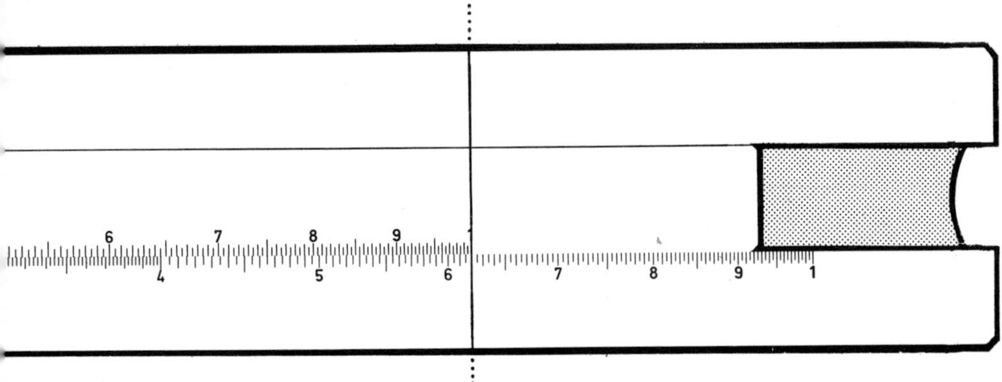

D-62

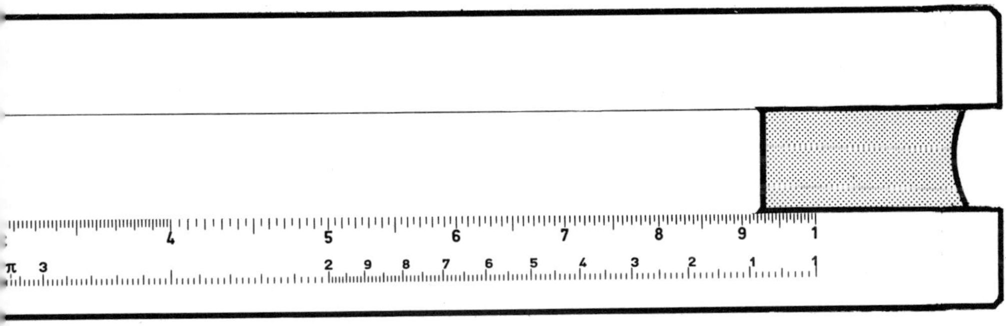

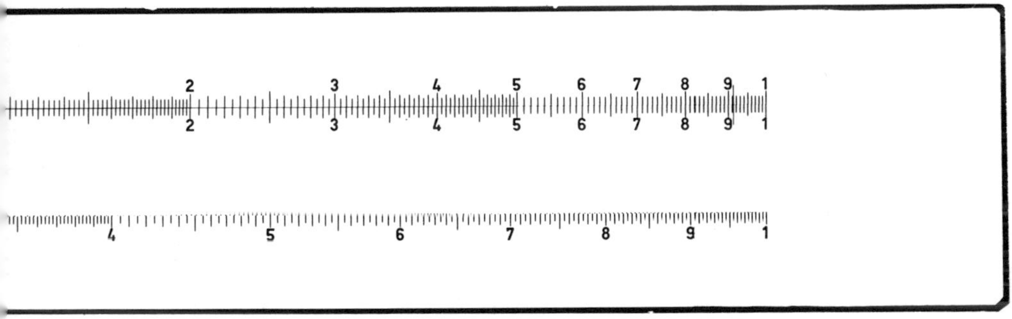

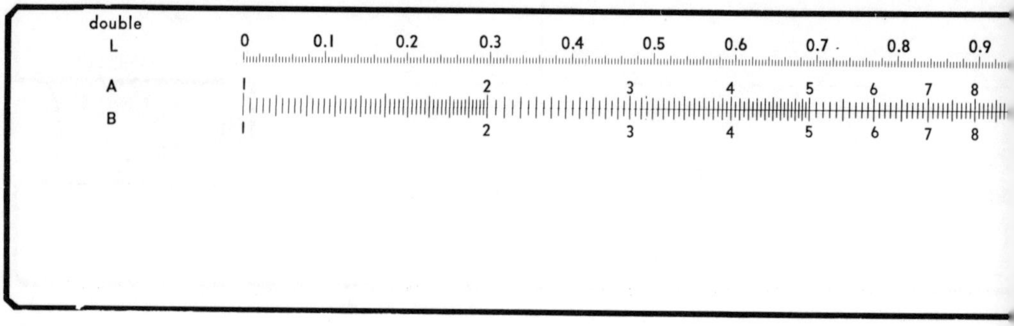

Figure 66

rithms matching the logarithmic readings of the L-scale (see page 74).

The L-scale runs from 0.0 to 1.0, but what if we were to construct such a scale with the units spaced more closely together, so that there would be room for a stretch from 0.0 to 2.0. If antilogarithms were lined up with such a "double L-scale," the results, as shown in Figure 66, would be a logarithmic scale running from 1 to 100.

Since the slide rule takes no account of the decimal point, the primaries of the A- and B-scales are not marked 1, 2, 3, 4, 5, 6, 7, 8, 9, 10, 20, 30, 40, 50, 60, 70, 80, 90, 100, but 1, 2, 3, 4, 5, 6, 7, 8, 9, 1, 2, 3, 4, 5, 6, 7, 8, 9, 1. There is thus an A-1, an A-1-(middle), and an A-1(right.)

On the whole it is best to let the A(left)-scale represent the stretch of the numbers from 1 to 10 and the A(right)-scale from 10 to 100, varying the decimal point two places at a time to get other stretches. Thus, the A(left)-scale includes not only, let us say, the num-

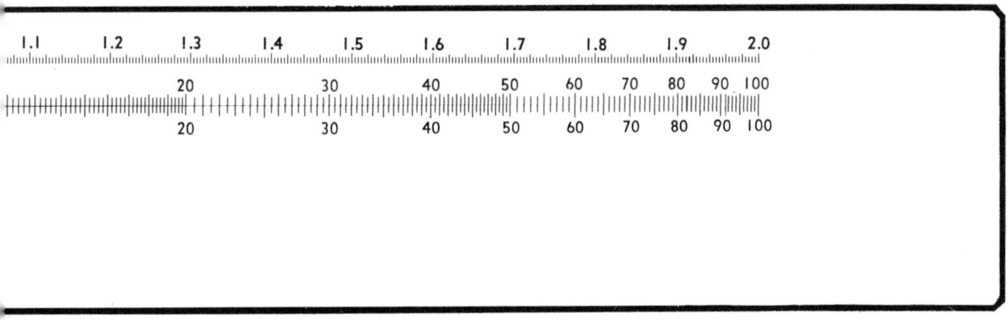

ber 5, but also 500; 50,000; 5,000,000; and so on. Working in the other direction, it also stands for 0.05, 0.0005, 0.000005, and so on. On the other hand, the A(right)-scale, includes 50; 5,000; 500,000; 50,000,000; and so on; as well as 0.5, 0.005, 0.00005, and so on.

This is not really difficult to remember. You will notice that the A(left)-scale includes the numbers with an even number of zeros neighboring the decimal point when written in the form given above. Thus, 500 has two zeros neighboring the decimal point, as has 0.05; and both 50,000 and 0.0005 have four zeros neighboring the decimal point. (Naturally, numbers less than one have to be written with the preliminary zero, as 0.05 and not as .05 if this rule is to hold.) The number 5 has no zeros and if this is considered an even number of zeros, this, too, fits the rule.

On the other hand, the A(right)-scale includes the numbers with an odd number of zeros neighboring the decimal point. Thus, 50 and 0.5, both have one zero neighboring the decimal point, 5000 and 0.005 both have three, and so on. All this, of course, holds for the B-scale as well.

The A(left)-scale crowds the full stretch of values found in the C- and D-scales into half the usual length. This means that the A(left)-scale cannot be subdivided as finely as the C- or D-scale (or any of the others of that family). There isn't the room.

The space between the numbered primaries of the A(left)-scale are divided into secondaries representing tenths, which are spaced more closely, of course, than are the secondaries in the C- and D-scales. Even between the primaries 1 and 2, where there is the most room, the secondaries are crowded too closely together to allow a convenient marking off by number as is done in the C- and D-scales and the rest of that family. None of the secondaries of the A- and B-scales are numbered.

The space between the secondaries lying in the range from 1 to 2 on the A(left)-scale are split into five subdivisions marked off by tertiaries representing 0.02 each. From 2 to 5, the space between the secondaries are divided in half, with a single tertiary representing 0.05. Between 5 and 1, there are no tertiaries at all. The markings on the A(right)-scale and of course on the B(left)- and B(right)-scales are identical with those on the A(left)-scale.

If you make use of the A- and B-scales, it is important to keep a wary eye on their markings and make sure you do not carry over the habits developed on the C- and D-scales.

You can multiply and divide on the A- and B-scales precisely as you can on the C- and D-scales. There are both advantages and disadvantages to this. I will mention the advantages first.

Since there are two scales, a left and a right, included in the A- and B-scales, there is no likelihood of the slide moving off the scale and you do not have to interchange indices.

Suppose, for instance, you multiply 6 by 2 on the C- and D-scales and carelessly begin by putting C-1 over D-6. You must next move the hairline to C-2, which, however, is well beyond the right edge of the slide rule. Nor can you switch to the CF- and DF-scales where, in this case, CF-2 is even farther beyond the right edge. There is no recourse but to switch indices.

If you did the same thing on the A- and B-scales, you would have no difficulty. It is the B-scales that is on the slide, so it is that which you move. Place B(left)-1 under A(left)-6 and now move the indicator to B(left)-2. To be sure, B(left)-2 is beyond the right end of the A(left)-scale, but here we have an A(right)-scale to take up the slack (Figure 67). B(left)-2 is under A-(right)-12 so $6 \times 2 = 12$. It is for this reason that there is no need for a folded A- or B-scale.

The A- and B scales can be used for division, too, in a manner analogous to the C- and D-scales. Here, too, the divisor is best placed on the slide; that is, on the B-scale.

Consider the problem $5.4 \div 2.2$. The divisor is 2.2

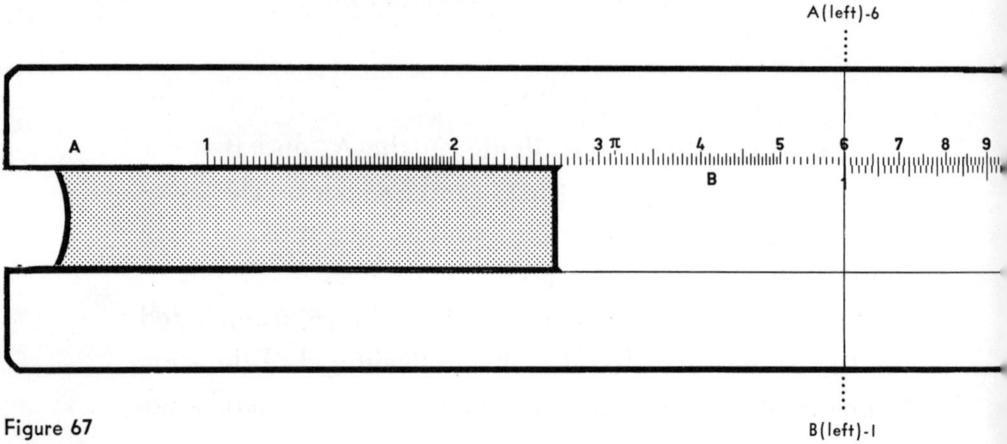

Figure 67

so you place it on the B-scale; B(left)-22 is put directly under A(left)-54 with the help of the hairline, and over B(left)-1 is A(left)-245 (Figure 68). Adjusting the decimal point, you decide that $5.4 \div 2.2 = 2.45$.

The A- and B-scales can also be used to find recipro-

Figure 68

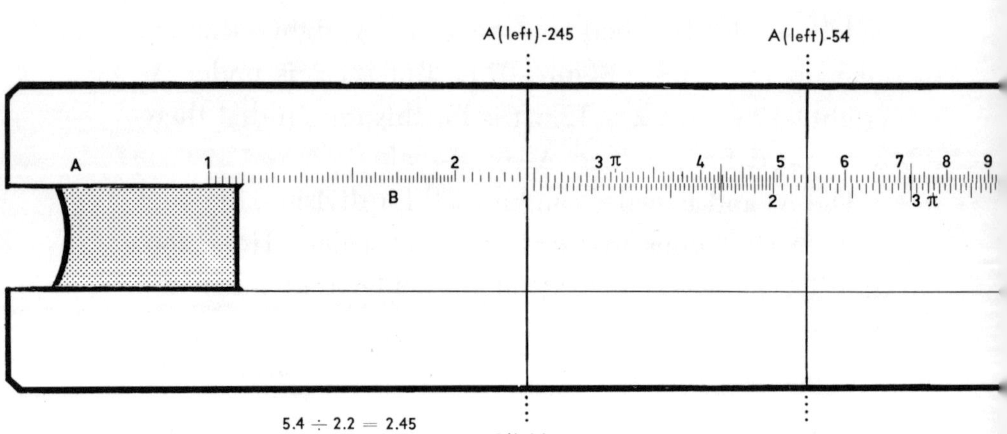

t)-12

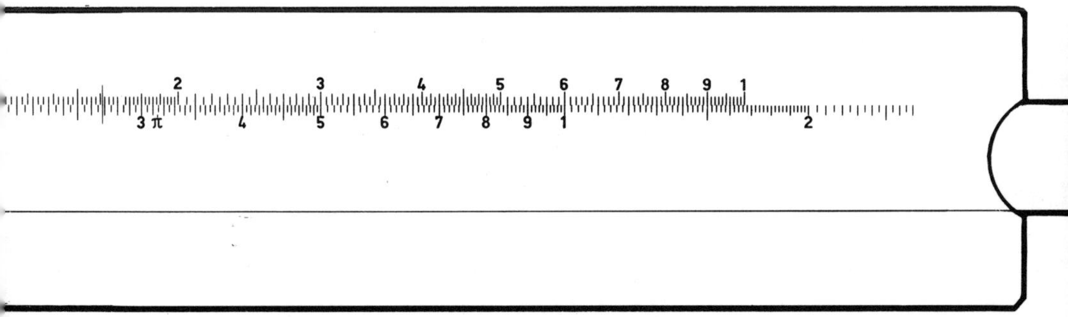

t-2)

cals in just the same way that the C- and D-scales can be used (see page 129).

Squares

The A- and B-scales can do everything, it seems, that

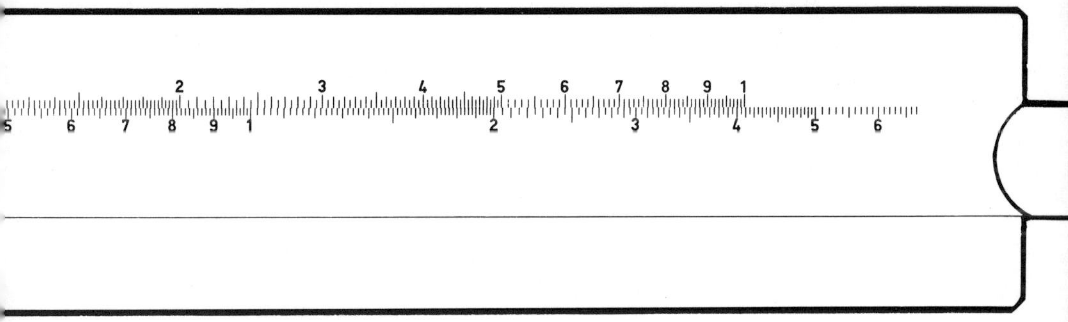

the C- and D-scales can do, without the danger of having to switch indices in mid-problem. Nevertheless, there is a serious disadvantage to their use.

Since the markings on the A- and B-scales are only half as far apart as the equivalent markings on the C- and D-scales, the A- and B-scales cannot be read as closely. This means that any answers that you get on the A- and B-scales are not as accurate as those you get on the C- and D-scales.

This question of accuracy is so important that ordinary multiplication and division should always be done on the C- and D-scales and never on the A- and B-scales. Increasing the accuracy is so desirable that one should willingly pay the price of an occasional shift of indices, or of having to transfer attention to the folded scales.

Yet I am not suggesting that the A- and B-scales be completely ignored, that they never be referred to. The A-scale, at least, has a particular convenience that is quite sufficient to justify its existence.

To see what that is, let me remind you again that the C- and D- scales are built up on the L-scale, which runs the length of the slide rule with a range from 0.0 to 1.0; while the A- and B-scales are built up on a similar "double scale" with a range from 0.0 to 2.0 (see page 157).

Since the double scale covers twice the range in the same distance, it moves, so to speak, twice as fast. At a given point where the L-reading is, let us say *a*, the "double L"-reading would be 2*a*.

Let us now suppose that the hairline is over a particular D(back)-reading and is simultaneously over a particular A-reading; we can call them D(back)-x and A-y. The distance of D(back)-x from D(back)-1 represents the logarithm of x and the distance of A-y from A(left)-1 represents the logarithm of y on the double scale. It is the same distance in both cases, but the double scale logarithm is twice the size of the ordinary logarithm at the same point, so that we can say $\log y = 2 \log x$, or $\log y = \log x + \log x$ (Figure 69).

If we shift to antilogarithms, remembering to convert an addition into a multiplication, we have $y = x \times x$, or to use the usual mathematical terminology, $y = x^2$.

What we have now shown is that when the hairline marks off simultaneous readings on the D(back)-scale and the A-scale, the A-reading is the square of the D(back)-reading.

To determine a square, then, we need only adjust the hairline without moving the slide. To determine the square of 5.1, place the hairline over D(back)-51 and note that it also gives a reading of A(right)-26. We can establish the order of magnitude by remembering that 5.1 is not much different from 5 and that the square of 5 is 25. Therefore $5.1^2 = 26$.

In the same way, D(back)-384 corresponds to A(right)-147 (Figure 70). To place the decimal point, we substitute 4 for 3.84. Since we know that $4^2 = 16$, that $40^2 = 1600$ and that $0.4^2 = 0.16$, we can tell that $3.84^2 = 14.7$, that $38.4^2 = 1470$, and that $0.384^2 = 0.147$.

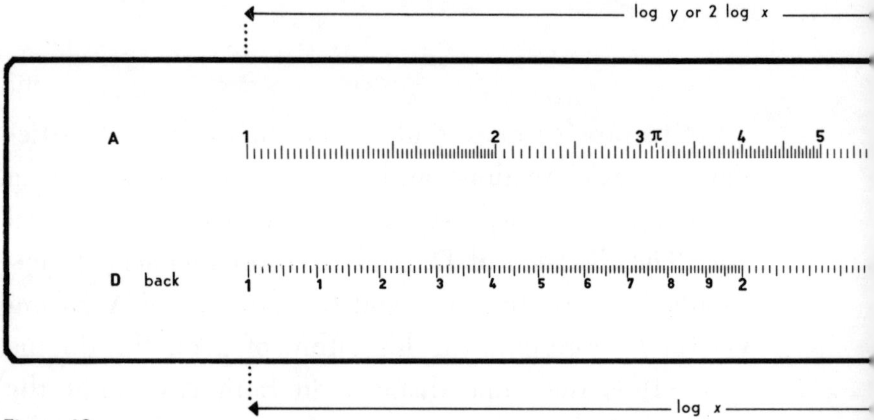

Figure 69

It is helpful, in connection with finding squares and placing the decimal point, to remember the odd-even rule concerning the right and left halves of the A-scale. The A(right)-scale involves numbers with an odd number of zeros neighboring the decimal point. For that reason, you may expect its readings to be of the order

Figure 70

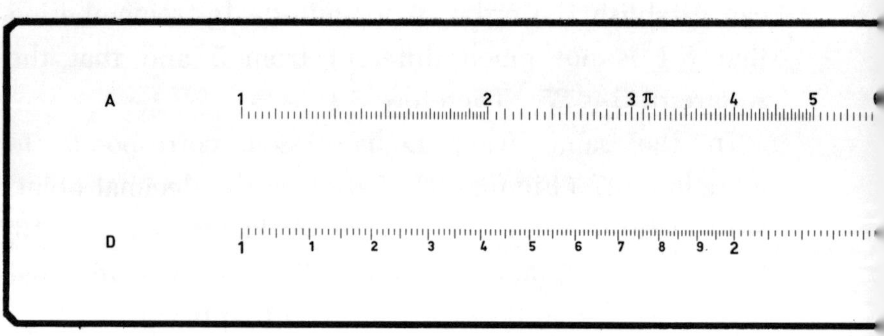

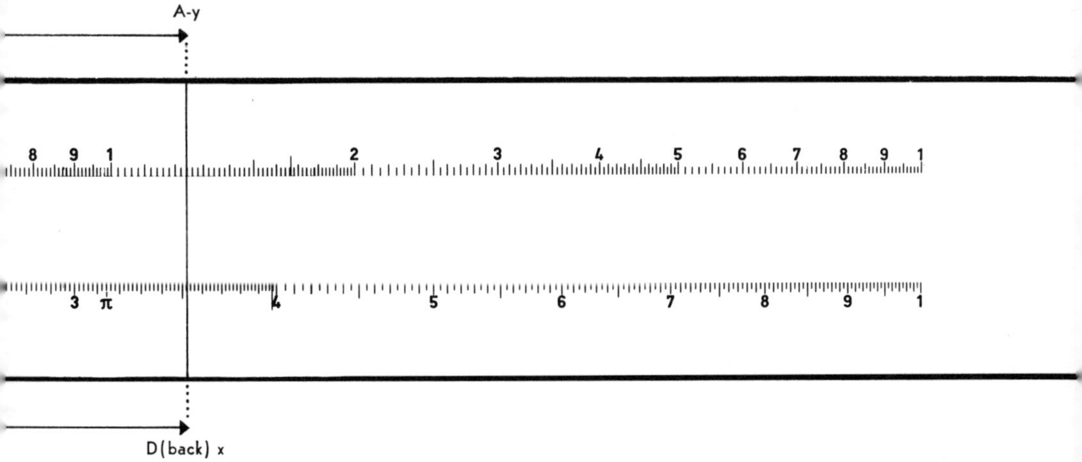

of magnitude, let us say, of 10 or 1000, but never 100. Therefore since the squares of 384 and of 51 fall in the A(right) range, their squares might be 14.7 or 1470 in the first case or 26 or 2600 in the second. Never, no matter how you place the decimal point in 384 or 51, can the square turn out to be 1.47 or 147 or 2.6 or 260.

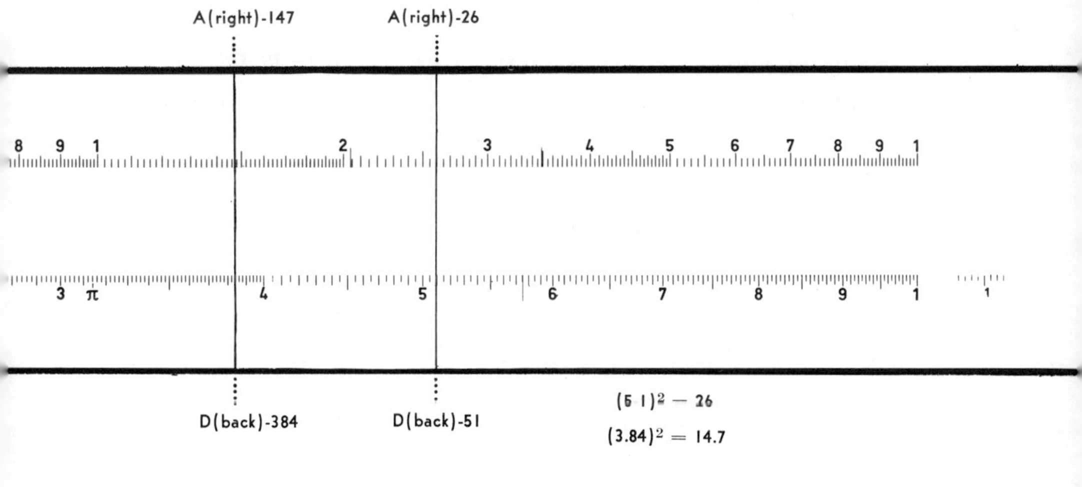

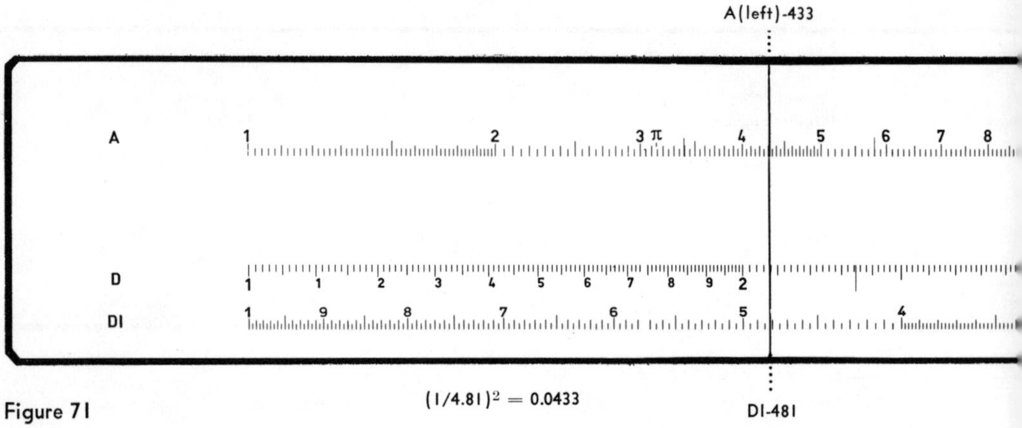

Figure 71

Another point to remember is that the D(back)-scale is not an absolute necessity for determining squares on the A-scale. The D-scale itself can also be used. Once a D-reading is set, one need only flip the slide rule to the other side to find the squares marked off as a simultaneous A-reading.

To be sure, squares can be determined without the A-scale too. If 3.84^2 is considered to be 3.84×3.84, that multiplication can be carried through on the C- and D-scales in the ordinary way. If this is done, the answer is obtained with greater accuracy, because the final reading is made on the D-scale and not on the less finely divided A-scale.

If, indeed, we carry through 3.84×3.84 on the C- and D-scales by putting C-1(right) over D-384, then moving the hairline to C-384, we will find, under it, D-1475. We will conclude that $3.84^2 = 14.75$. This is

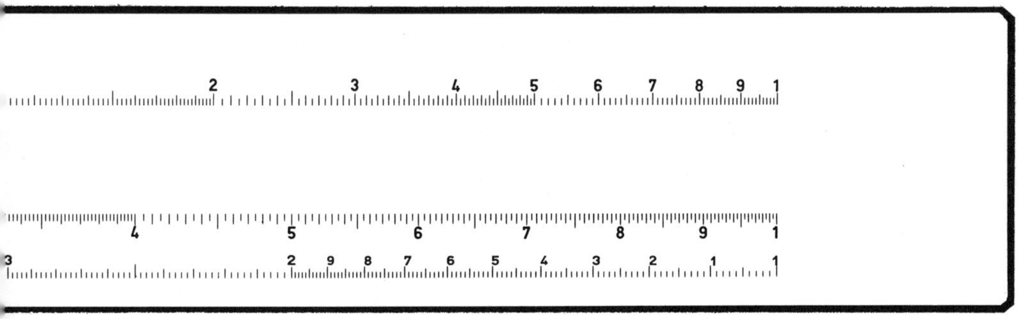

closer to the truth than the $3.84^2 = 14.7$ determined by use of the D- and A-scales; for the true answer, worked out in full, is 14.7456.

Nevertheless, working out the square by use of the A-scale involves no motion of the slide, and for quick work that maneuver comes in handy.

By substituting the DI-scale for the D(back)-scale, we can get the square of a reciprocal: say $\left(\frac{1}{4.8}\right)^2$. We can set the hairline on DI-48 and obtain a simultaneous reading of D(back)-208, which gives us the reciprocal $\frac{1}{4.8}$. However, we needn't stop to work out the decimal place or even look at the digit combination. We pass directly on to the A-scale, which will give us the square of the D(back)-reading and hence the square of the reciprocal of the DI reading.

We find the A-reading to be A(left)-433 (Figure 71),

and now we are ready for the decimal point. In place of 4.8, let us take 5. We therefore have $(1/5)^2$ which equals $(0.2)^2$ or 0.04, an order of magnitude with two zeros neighboring the decimal point (checking the fact that we ended with an A(left)-reading, which requires an even number of zeros). We conclude then that $\left(\frac{1}{4.8}\right)^2 = 0.0433$.

The fact that the slide rule gives us a method for quickly determining squares means that we can carry through a multiplication or a division and find the square of the result at once.

If there were a C(back)-scale adjoining the D(back)-scale on the reverse side of the slide rule (as there is, in some slide rule designs), we could run the multiplication or division on those scales, get the product or quotient on the D(back)-scale and take the simultaneous A-reading.

On my slide rule, however, there is no C(back)-scale. The reverse of the slide, where such a scale should be, is given over, instead, to several scales involving trigonometric functions which I will not discuss in this book.

Fortunately, this introduces no great difficulty. We can use the C- and D-scales on the face of the slide rule instead.

Suppose, for instance, we want to solve the problem $(9.44 \times 2.75)^2$. Beginning at the front of the slide rule, we place C-1(right) over D-944, move the hairline to C-275, and find a D-reading under it that represents the product of 9.44×2.75. However, we are not interested in the product but in its square, so we needn't even bother looking at the D-reading. Instead, we flip the slide rule in search of the A-reading and find it to be A(left)-673.

We get the order of magnitude by substituting 10 for 9.44 and 3 for 2.75. The problem becomes $(10 \times 3)^2 = 30^2 = 900$, which checks with the fact that an A(left)-reading requires an even number of zeros in the neighborhood of the decimal point. We conclude, then, that $(9.44 \times 2.75)^2 = 673$.

You can use the same principle to solve $(6.3 \div 1.18)^2$, provided you remember to manipulate the slide rule in such a fashion as to let the quotient of $6.3 \div 1.18$ fall upon the D-scale. Only if it falls upon the D-scale can the square of the quotient simultaneously be found on the A-scale. However, if we follow the rule of placing the divisor on the C-scale (see page 128), this point will take care of itself.

Since 1.18 is the divisor, it goes on the C-scale. Place C-118 over D-63. Move the hairline to C-1 and under it is D-535, which represents the quotient. Without bothering with the D-reading at all, however, you flip the slide rule and find A(right)-285. To place the decimal point you replace 6.3 by 6 and 1.18 by 1. The problem becomes $(6 \div 1)^2 = 6^2 = 36$. This order of magnitude is in the 10's, which checks with the fact that the final reading was found on the A(right)-scale, requiring an odd number of zeros in the neighborhood

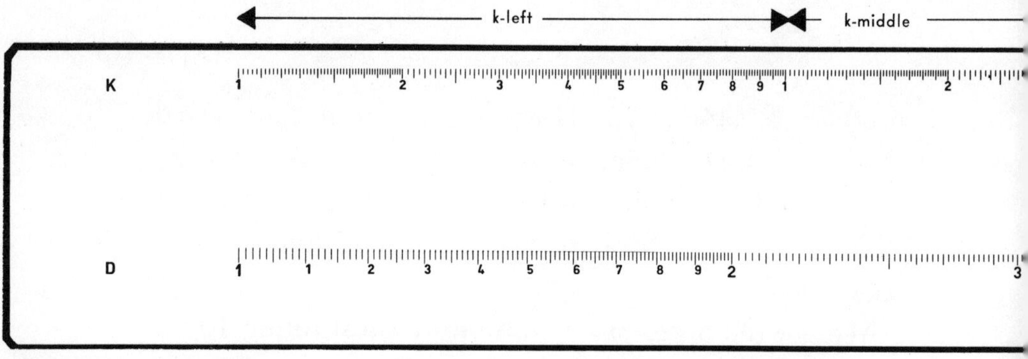

Figure 72

of the decimal point. Therefore $(6.3 \div 1.18)^2 = 28.5$.

Cubes

An exponential number like x^2 is said to be a *power of* x. The higher the value of the exponent, the higher the power. The average slide rule can reach a higher power than the square; it can, in fact, handle the *cube* (x^3) just as easily.

The square involved a double L-scale so it is not unexpected that the cube will involve a "triple L-scale." Since the ordinary L-scale covers the range of logarithms from 0.0 to 1.0, the triple L-scale will cover the range from 0.0 to 3.0. In terms of antilogarithms this is the range from 1 to 1000.

The antilogarithm scale with markings from 1 to 1000 is present on the front of my slide rule, all the way at the top, and is the K-scale (Figure 72). (In another slide rule I have, it is on the back, all the way on the bottom.) If you follow its markings you will see that its

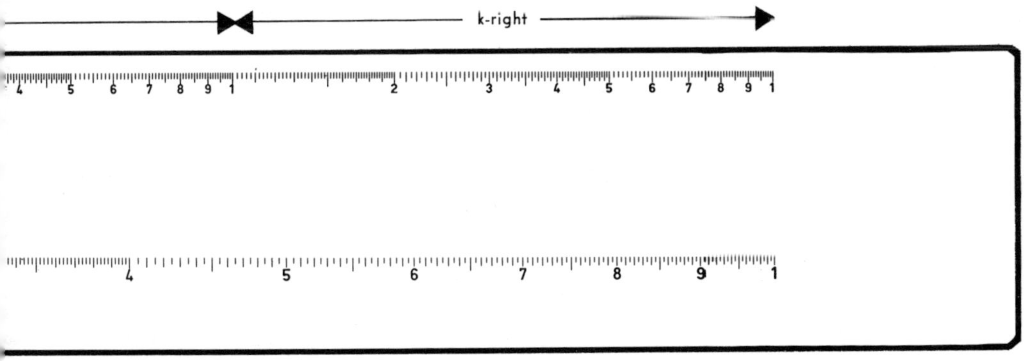

primaries represent 1, 2, 3, 4, 5, 6, 7, 8, 9, 10, 20, 30, 40, 50, 60, 70, 80, 90, 100, 200, 300, 400, 500, 600, 700, 800, 900, 1000. Since the decimal point is arranged to suit ourselves, it is only necessary to write the digits: 1, 2, 3, 4, 5, 6, 7, 8, 9, 1, 2, 3, 4, 5, 6, 7, 8, 9, 1, 2, 3, 4, 5, 6, 7, 8, 9, 1.

Just as the A- and B-scales are divided into two identical halves, the K-scale is divided into identical thirds, so that there is a *K(left)-scale,* a *K(middle)-scale,* and a *K(right)-scale.*

The decimal point jumps three places at a time in each of the three sections, thus:

K(left)-scale	0.002	2	2,000	2,000,000	
K(middle)-scale	0.02	20	20,000	20,000,000	
K(right)-scale	0.2	200	200,000	200,000,000	etc.

The individual thirds of the K-scale are even more crowded than the halves of the A- and B-scales. The spaces between 1 and 3 are marked off by secondaries with values of 0.1 and tertiaries with values of 0.05. The spaces between 3 and 6 are marked off by secondaries only, with values of 0.1 each; and the spaces between 6 and 1 by secondaries with values of 0.2.

The K-scale exists by itself; there is no similar scale that can be moved along it as the B-scale can move along the A-scale, or the C-scale along the D-scale. For that reason there is no way of using the K-scale directly for ordinary multiplications and divisions, and that is just as well, for the K-scale would be extremely inaccurate for that purpose.

The important use for the K-scale is in connection with cubes. In fact, the "K" of the K-scale stands for *Kubus,* the German word for cube.

The manner in which it was shown that the readings on the A-scale represent the squares of the corresponding readings of the D(back)-scale (see page 167) can be used (with appropriate modification for a triple L-scale, rather than a double) to show that K-readings represent the cubes of corresponding D(back)-readings.

For instance D-2 corresponds to K(left)-8; D-43 corresponds to K(middle)-8 and D-928 corresponds to K(right)-8. Remembering the discussion on decimal points above, we see that $2^3 = 8$, $4.3^3 = 80$, and $9.28^3 = 800$.

The method for solving problems such as $(64 \times 0.4)^3$ and $(3.11 \div 2.6)^3$ is just the same as for solving analogous problems involving squares (see page 167). We merely look for the K-reading, rather than the A-reading.

8

Roots

Square Roots

When we dealt with addition, we found we could also deal with its reverse — subtraction. Again, in dealing with multiplication, we dealt with its reverse — division. Now, in connection with powers, we find that the reverse operation can also be dealt with, and it involves roots.

The square root of a is that number which, when multiplied by itself, yields a as a product. In other words if $b \times b = a$ (an equation which can also be written $b^2 = a$), then b is the square root of a, and this can be written $b = \sqrt{a}$. The square and the square root are thus opposite sides of a coin, so to speak. If a is the square of b, then b is the square root of a.

We can make this plainer with numbers, perhaps. Since $5 \times 5 = 25$, 25 is the square of 5, and 5 is the square root of 25. Since $6 \times 6 = 36$, then 36 is the square of 6, and 6 is the square root of 36.

To find the square of a number, using the C- and D-scale, as I pointed out earlier (page 167), is simple enough. To determine 7.8^2, you need only solve for

7.8×7.8 in the usual way and find the answer to be 60.8.

To find the square root of a number using the C- and D-scale is quite another thing. Suppose you want the solution to $\sqrt{7.8}$. That means you want a number which, when multiplied by itself, will yield 7.8 as a product. You find that product on the D-scale and mark D-78 with the hairline.

Next you must adjust the slide in such a way that the C-reading at the hairline over D-78 must be the same as the D-reading under C-1. By careful manipulation of the slide you will find that when C-279 is over D-78, C-1 is over D-279. That means that $2.79 \times 2.79 = 7.8$ and, therefore, that $\sqrt{7.8} = 2.79$.

Adjusting the slide to find a square root is not an impossible procedure, but it is a tedious one, and must be carried through carefully, with frequent glances back and forth between C-1 and the hairline. It is not at all the usual flick-flick of the slide rule as in carrying through so many other computations.

Consequently, if another technique will suffice to find the square root, it should be used.

The answer lies with the A-scale and with the knowledge that finding a root is the reverse of the operation of finding a power. Since the A-reading is the square of the simultaneous D(back)-reading, the D(back)-reading is the square root of the simultaneous A-reading.

To find the square root of 7.8, then, it is only necessary to place the indicator on A(left)-78. The simultaneous reading of D(back)-279 (Figure 74) is sufficient to tell us that $\sqrt{7.8} = 2.79$, with that single setting of the indicator. Not only is it not necessary to adjust the slide carefully back and forth to find the square root, it isn't necessary to move the slide at all.

By combining the A- and DI-scales, we can obtain the reciprocal of the square root. If we wanted $\frac{1}{\sqrt{7.8}}$, we would set the hairline at A(left)-78 and note the simultaneous reading of DI-358 (Figure 73). We therefore conclude that $\frac{1}{\sqrt{7.8}} = 0.358$. (To locate the decimal point, we need only replace 7.8 by the fairly close approximation 9. We then see that $\frac{1}{\sqrt{9}} = \frac{1}{3} = 0.333$.)

Notice that we determined the square root of 7.8 to be 2.79 by setting the hairline at A(left)-78. What if we had set the hairline at A(right)-78? If we do so we would find a simultaneous reading of D-878. The difference is a matter of decimal point. We are now dealing not with 7.8 but with 78, and we can say that $\sqrt{7.8} = 8.78$.

In finding a square root, we must first determine whether to use the A(left)-scale or the A(right)-scale. To do that we deal only with the first digit of the number whose square root we want, converting the other digits into zero. We then count the number of zeros in the neighborhood of the decimal point, and for even numbers use the A(left)-scale, and for odd numbers the

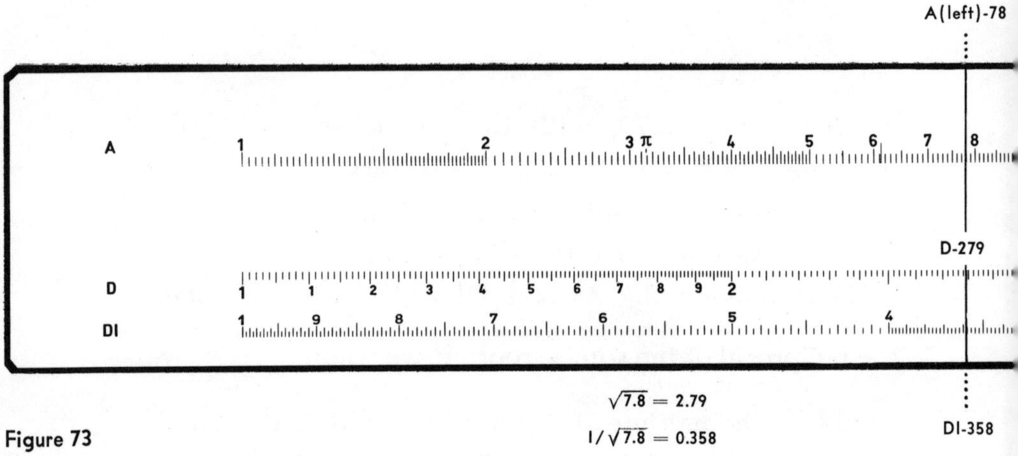

Figure 73

A(right)-scale, in accordance with the rule given on page 161.

For the square root of 723, we use the number 700 as a guide. Since there are an even number of zeros, we use the A(left)-scale. For the square root of 0.562, we use 0.5 as the guide. Now we have an odd number of zeros and use the A(right)-scale.

If your memory fails you, you can always use the method of approximation. Suppose that you want the square root of 67. If you use the A(left)-67 setting, you obtain a reading of D(back)-259. On the other hand, if you use the A(right)-67 setting, you obtain a reading of D(back)-819. You need not hesitate between the two. Since 67 is not far removed from 64, and you know that the square root of 64 is 8 (since $8 \times 8 = 64$), you have no need to pause in saying that $\sqrt{67} = 8.19$.

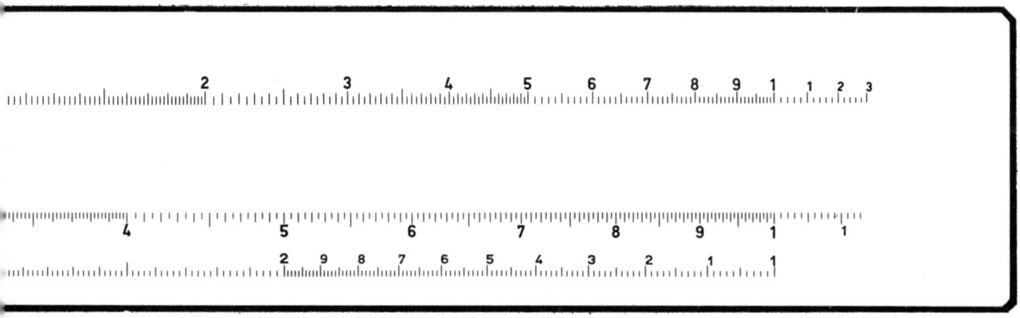

(On the other hand, $\sqrt{6.7} = 2.59$.) Of course, if you had treated 67 as 60 and noted the odd number of zeros — just one, that is — in the neighborhood of the decimal point, you would have used the A(right)-scale at once.

Cube Roots

Just as we could reverse the procedure of finding squares in order to make it possible for us to find square roots, so we can reverse the procedure of finding cubes so as to make it possible for us to find *cube roots.*

The cube root of b is the number which, when multiplied by itself twice, gives b as a product. If $a \times a \times a = b$ (which can also be written $a^3 = b$), then a is the cube root of b, or, as it is usually written, $a = \sqrt[3]{b}$. Cubes and cube roots are inverses of each other. If b is the cube of a, then a is the cube root of b.

We already know that if we simultaneously take a K-reading and a D-reading, the K-reading is the cube of the D-reading (see page 176). It follows, therefore, that the D-reading is the cube root of the K-reading.

Suppose we wanted $\sqrt[3]{7.2}$. Place the hairline at

Figure 74

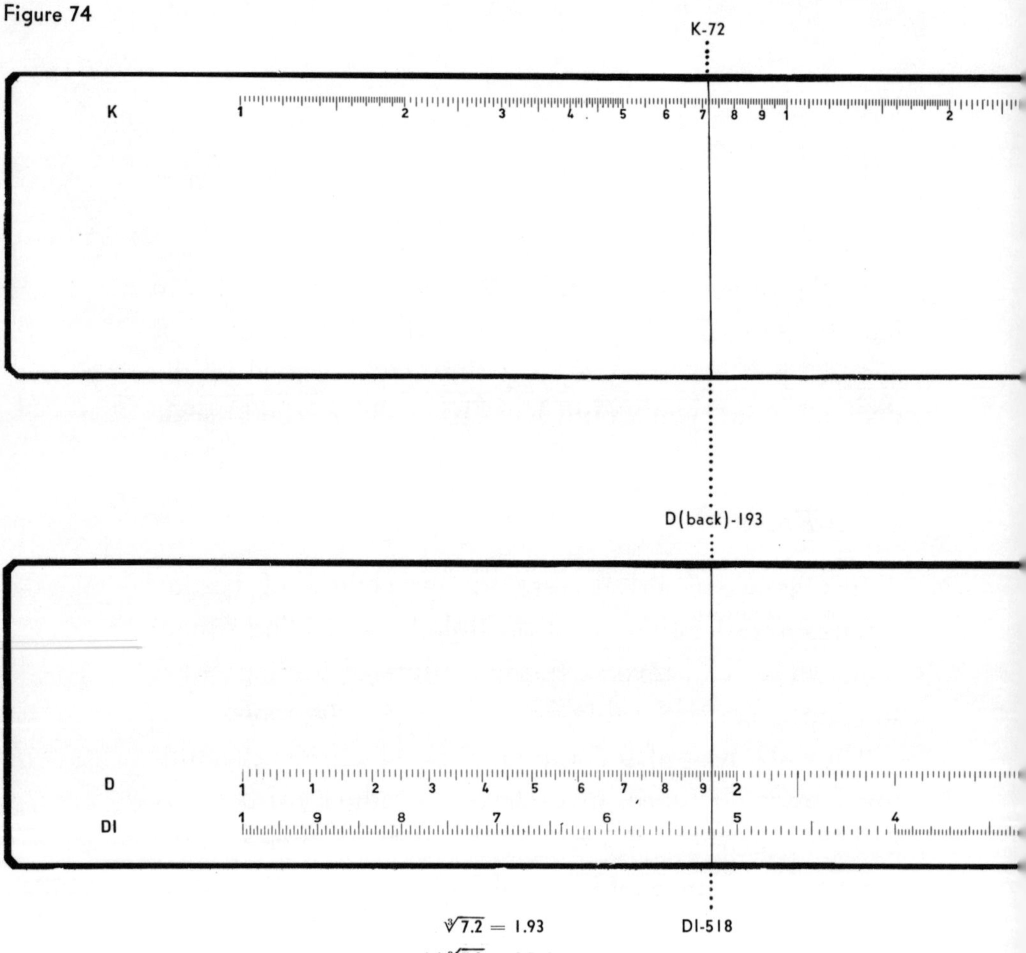

K(left)-72 and the simultaneous reading of D-193 tells us that $\sqrt[3]{7.2} = 1.93$ (Figure 74). The decimal point is located by the fact that if we replace 7.2 by 8, we know at once that the $\sqrt[3]{8} = 2$, since $2^3 = 8$.

By combining the K- and DI-scales, we can obtain the reciprocal of the cube root. Since K(left)-72 gives a

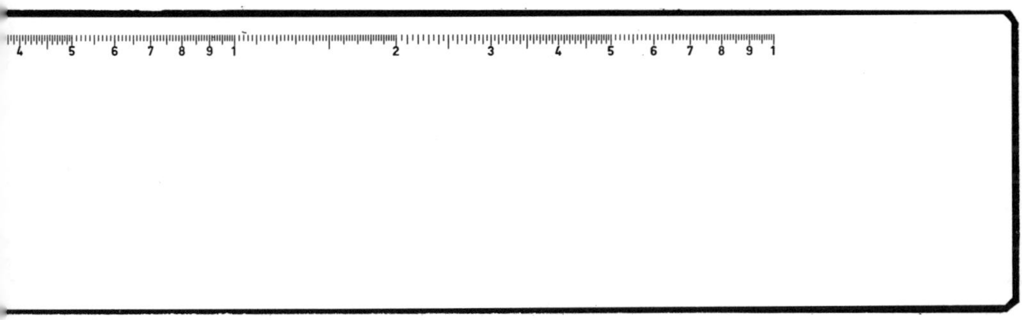

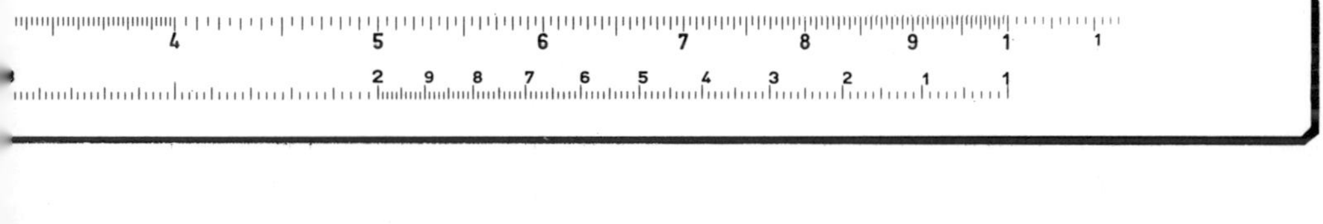

simultaneous reading of DI-518, we know that $\frac{1}{\sqrt[3]{7.2}} =$ 0.518. Again, we locate the decimal point by replacing 7.2 with 8. Since $\frac{1}{\sqrt[3]{8}} = \frac{1}{2} = 0.5$, we have our answer.

In finding cube roots, however, we have three K-scales to choose from for our setting. In finding the cube root of 7.2, for instance, do we use K(left)-72, K(middle)-72, or K(right)-72? The answer lies in the table on page 175. For numbers from 1 to 10, we use the K(left)-scale, between 10 and 100 the K(middle)-scale, between 100 and 1000 the K(right)-scale. For numbers from 1000 to 10,000, we are back to the K(left)-scale, and so we progress through the three scales in order indefinitely.

The same holds true for numbers less than 1. From 1000 down to 100 we use the K(right)-scale, as I have already said, from 100 down to 10 the K(middle)-scale, and from 10 down to 1 the K(left)-scale. Progressing then in this direction, we use the K(right)-scale for numbers from 1 down to 0.1, the K(middle)-scale for 0.1 down to 0.01 and so on.

* * * * *

In this book I have dealt with twelve scales on the slide rule. With those twelve scales, we can multiply, divide, take proportions and reciprocals, obtain squares, square roots, cubes, cube roots, logarithms, and antilogarithms.

Such calculations (together with additions and subtractions for which we don't need slide rules) probably take care of virtually all the calculations that we encounter in ordinary life.

There are, to be sure, other types of computations that can be performed on slide rules. Those scales on my slide rule which I haven't discussed in this book make it possible to deal with fractional exponents, with natural logarithms, with trigonometric functions, and so on.

Then, too, special scales not found on ordinary slide rules are adapted to the particular needs of those who must do calculations in special fields, such as those of electrical engineering or of business. By bending the slide rule into circles or spirals, a greater length of scale (and hence greater accuracy) can be squeezed into a particular space.

What we have described in this book, therefore, by no means exhausts the usefulness of the slide rule.

But we have made a good beginning. Armed with the slide rule, there will be few ordinary problems that need hold any terrors for you. A push of the slide this way and that, an adjustment of the hairline, and the answer is yours!

No wonder that to many people the feel of the slide rule in the hands spells security.

As I said in the introduction — a man who must carry out numerical computations constantly would be as lost without his slide rule as a doctor without his stethoscope or a painter without his brush.

INDEX